Agent-Based Social Systems

Volume 15

This series is intended to further the creation of the science of agent-based social systems, a field that is establishing itself as a transdisciplinary and cross-cultural science. The series will cover a broad spectrum of sciences, such as social systems theory, sociology, business administration, management information science, organization science, computational mathematical organization theory, economics, evolutionary economics, international political science, jurisprudence, policy science, socioinformation studies, cognitive science, artificial intelligence, complex adaptive systems theory, philosophy of science, and other related disciplines.

The series will provide a systematic study of the various new cross-cultural arenas of the human sciences. Such an approach has been successfully tried several times in the history of the modern science of humanities and systems and has helped to create such important conceptual frameworks and theories as cybernetics, synergetics, general systems theory, cognitive science, and complex adaptive systems.

We want to create a conceptual framework and design theory for socioeconomic systems of the twenty-first century in a cross-cultural and transdisciplinary context. For this purpose we plan to take an agent-based approach. Developed over the last decade, agent-based modeling is a new trend within the social sciences and is a child of the modern sciences of humanities and systems. In this series the term "agent-based" is used across a broad spectrum that includes not only the classical usage of the normative and rational agent but also an interpretive and subjective agent. We seek the antinomy of the macro and micro, subjective and rational, functional and structural, bottom-up and top-down, global and local, and structure and agency within the social sciences. Agent-based modeling includes both sides of these opposites. "Agent" is our grounding for modeling; simulation, theory, and realworld grounding are also required.

As an approach, agent-based simulation is an important tool for the new experimental fields of the social sciences; it can be used to provide explanations and decision support for real-world problems, and its theories include both conceptual and mathematical ones. A conceptual approach is vital for creating new frameworks of the worldview, and the mathematical approach is essential to clarify the logical structure of any new framework or model. Exploration of several different ways of real-world grounding is required for this approach. Other issues to be considered in the series include the systems design of this century's global and local socioeconomic systems.

More information about this series at http://www.springer.com/series/7188

Utomo Sarjono Putro • Manabu Ichikawa
Manahan Siallagan
Editors

Agent-Based Approaches in Economics and Social Complex Systems IX

Post-Proceedings of The AESCS International Workshop 2015

Springer

Editors
Utomo Sarjono Putro
School of Business and Management
Institut Teknologi Bandung
Bandung, Indonesia

Manabu Ichikawa
National Institute of Public Health
Wako, Saitama, Japan

Manahan Siallagan
School of Business and Management
Institut Teknologi Bandung
Bandung, Indonesia

ISSN 1861-0803
Agent-Based Social Systems
ISBN 978-981-10-3661-3 ISBN 978-981-10-3662-0 (eBook)
DOI 10.1007/978-981-10-3662-0

Library of Congress Control Number: 2017937543

Printed on acid-free paper

This Springer imprint is published by Springer Nature
The registered company is Springer Nature Singapore Pte Ltd.
The registered company address is: 152 Beach Road, #21-01/04 Gateway East, Singapore 189721, Singapore

Contents

Understanding Risks in Milk Supply Chain: An Agent-Based Conceptual Model

Andre R. Daud and Utomo S. Putro

Abstract Milk production is inherently complex and risky business. Risks and complexities are involved in every production stage and process performed by each actors involved along the milk supply chain. It implies that the success of improving the performance of supply chain will in some degree depend on those actors' ability to cope with risks and its emerging complexities. This paper proposes a concept for analyzing the risks existing in the milk supply chain. To conceptualize, the use of agent-based methodology is offered in order to capture the risk complexities in the particular chain. A basic agent-based model that can describe the actors' behavior toward risk and the likely consequences for the entire chain completes the concept.

Keywords Agriculture • Behaviors • Complexity • Dairy production

1 Introduction

Risk in agricultural production has been extensively studied since the beginning of modern agricultural sector. In this phase, risk was being the main concern upon the society, as well as scholar and government, because it may often associate with adversity and loss by the agricultural firm and also with its survival as a business. As rural family farms initially dominate the agricultural sector, most risk studies emphasize on the survivability of their business which being the main source for income and welfare of rural community. Concisely, risk is prevalent in the agricultural operations.

One definition of risk is as uncertainty that "matter" for producers and may involve the probability of losing money or welfare [1, 2]. Previous studies have revealed many sources of agricultural risk, and there are evidences showing risk-aversion as most typically agricultural producers' attitude toward risk. Risk averse

A.R. Daud (✉)
Faculty of Animal Husbandry, Universitas Padjadjaran, Bandung, Indonesia
e-mail: andre.daud@sbm-itb.ac.id

U.S. Putro
School of Business and Management, Institut Teknologi Bandung, Bandung, Indonesia

U.S. Putro et al. (eds.), *Agent-Based Approaches in Economics and Social Complex Systems IX*, Agent-Based Social Systems 15, DOI 10.1007/978-981-10-3662-0_1

farmers will prefer to abandon some potential outcome to avoid any possible future risks that may make them suffered from losing some amount of income. In this situation, the outcome of agricultural production would be less optimal under the presence of risks. Thus, producers seek to avoid risk through various managerial and institutional mechanisms.

However, despite from the advancement of studies, our understanding on risk and risk management strategies in agricultural production is still limited. Most of agricultural risks and its associated management are still positioned only in the standpoint of individual entity (for example, in the individual farmers' context). This may not become relevant any further because agricultural production recently has transformed itself to a supply chain in which involves a set of collective entities that performed wide range of activities and processes, from farm-household producers in rural areas into large-modern food processors, retailers and markets in urban area [3, 4]. This condition implies the nature of strong interdependency among business actors, and in turn, complexities in managing risks emerge.

Given the characteristic, more comprehensive approach is considered necessary in risk studies for agricultural supply chain. Recently, agent-based model (ABM) methodology is preferred by scholars in many studies which focus on individuals or agents that are described as unique and autonomous entities that usually interact with each other and their local environment. Railsback and Grimm [5] suggest that ABM can provide a method to address problems that concern emergence complexity. The emergence is system dynamic that arises from how the system's individual components interact with and respond to each other and their environment. The possibility for ABM to address such a problem comes from the model's ability to work across level, i.e., working vice versa between individual and its system. Thus, ABM methodology seems quite feasible to be applied to supply chain risk studies, particularly in agricultural production.

This paper presents a conceptual framework for analyzing risk specifically in milk supply chain in the context of developing countries. Instead of classical approach, which mostly emphasizes on explaining the individual's attitude toward risks, ABM is proposed to be the main method for capturing the effect of risk on the supply chain actors collectively, as will be presented later. After this introduction section, an overview on the complexity in milk production is briefly presented. Thereafter, the framework that combines the concept of supply chain, risks, and modeling is proposed. In the end of this paper, a summary section will give the brief overview of the benefit of ABM approach to be applied in the case of milk supply chain.

2 Complexity in Milk Supply Chain

In agricultural research, the use of ABM approach to understand individuals and the whole system of agriculture is relatively new. Although agricultural field has its own approach to understand the system, i.e., biophysical approach, it is still limited

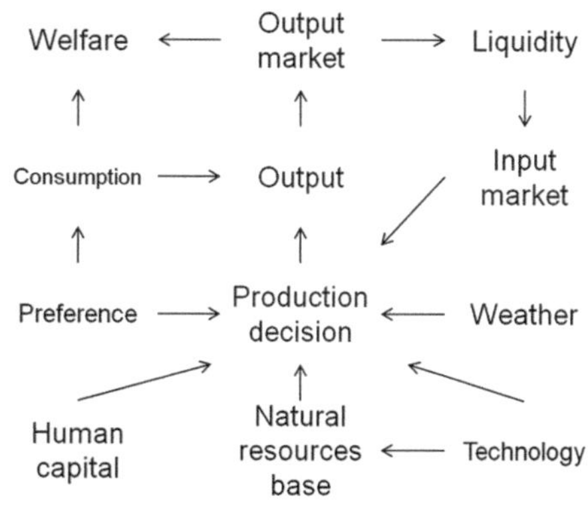

Fig. 1 Agricultural production system complexity (Adapted from Bebe et al. [6])

and constrained by existence of the complexity in the agricultural system. Indeed, agricultural production is one of the systems that are very complex. It is because of the agricultural production which has multidimensional aspect such as the entire biophysical socioeconomic system in the real world. To provide its complex system, we illustrate a generic form of agricultural production system in this figure (Fig. 1).

Agricultural production system, presents all of agents, interactions, and its environment. As we can observe, the agricultural production system comprises many agents who have many interactions within their agricultural environment. In traditional milk production system, millions of smallholder farmers – as the most important agent in the system – with their heterogeneous characteristics and attributes, have their own autonomy to decide how much dairy cow they should keep, to whom they should sell their milk products, on what price and on what market, thus avoiding risks, to satisfy their given family household objectives, under environment uncertainty. Similar situation applies for other agents in this milk production system, thus magnifying the complexity of the system.

As nowadays milk production is in the form of a supply chain, it has been regarded as "complicated systems involving both strategic and operational issues along with complex social and functional behaviors [7, 8]". Indeed, Choi et al. [9], Pathak et al. [10], and Surana et al. [11] argued the supply chain is "to be regarded as complex adaptive systems (CAS) because individual components or agents within a supply chain can and do intervene at any point in a meaningful way to change the behavior of the whole". Under these circumstances, an agent may represent an individual, a group, or an entire organization with each having relationship and varying degrees of connectivity with other agents. These allow information and resources to flow between agents.

A widely accepted description of complex adaptive systems is "a system that emerges over time into a coherent form, and adapts and organizes itself without any singular entity deliberately managing or controlling it" [9]. Pathak [10] presented several common features for complex adaptive systems, which are classified into two characteristics, micro and macro. Microlevel describes the basic structure of the system, i.e., the structure which is internally built by the presence of many agents,

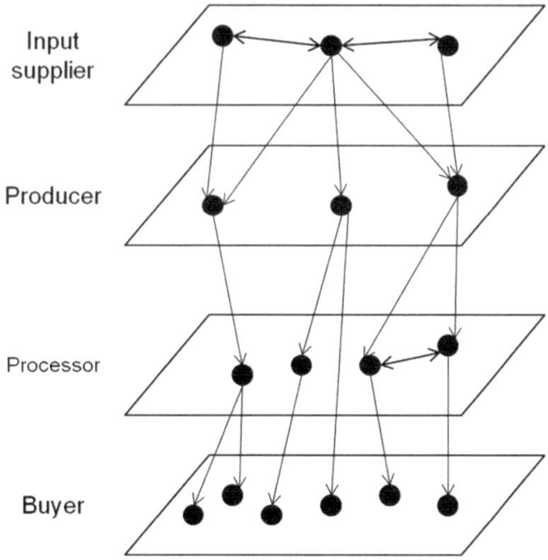

Fig. 2 Network supply chain/net chain (Adapted from Lazarrini [12])

while the macro-level describes how a whole system behaves. Following Lazarrini [12], the micro- and macro-level characteristics of complex adaptive systems in particular milk supply chain can be shown in the following illustration.

On the microlevel, the presence of a variety of actors is the main characteristic of CAS. As can be seen in Fig. 2, there are usually many actors in the supply chain – farmers, processor, manufacturers – who have diverse characteristics and different preferences in the terms of production technology, product features, and interactions, both vertically and horizontally. This implies that there are many different needs and objectives possessed and decision making performed by each actor. In the milk supply chain, this condition increases complexity of particular supply chain system. The presence of local interactions and interdependencies among system components is also the main source of complexity of a system which resulted from various actors. Surana et al. [11] showed that in a supply chain, there could be physical interaction and social as well. With the presence of many actors, these interactions can be numerous and heterogeneous, thus generating an interconnection complexity.

Nestedness and adaptiveness are other microlevel characteristics of CAS [11]. Nestedness refers to the behavior that is formed hierarchically. For example in milk supply chain, the behavior of dairy farmers and their interactions determine the behavior of each farmer group to which they belong. In turn, the behavior of dairy farmer groups and their each interaction internally and externally with another actors in the chain will define the behavior of a milk supply chain system as a whole. Closely, adaptiveness refers to the ability to change behaviors according to their system environment [15] as the behavior of actors can be influenced by its interactions with the environment [11]. For instances, the actual practices of keeping dairy cattle in suburban area will differ significantly with them practicing dairy in rural area.

For the characteristic in macro-level, there are emergent behavior, self-organization, path dependency, and coevolution as features in a complex system. Emergence refers to the behavior in a system that emerges from the behavior of individual components (both social and physical) and their interactions [15]. In the context of milk supply chain, especially in Indonesia, declining trend of milk production is evidenced. It follows the very low scale of dairy operation preferred by farmers. Actually, their behavior often arises without influences from external or central control in the system but from the results of many autonomous interactions between agents in the system.

The CAS also features coevolution and path dependency. Coevolution refers to changes in the structure of a particular system as results from learning and adaptation process experienced by agents in their interaction with the environment. In turn, the structural changes trigger the coevolution of the system and its environment. Following this dynamic, the actions and decisions made in previous state of the system will determine the current and future states. This is defined as path dependency of the system. Choi et al. [9] also showed that the options in the current state are the reflection of microlevel decision making made by actors in the past.

To assess the CAS, Macal and North [15] have suggested a typical of agent-based model (ABM) which comprises three basic elements: (1) a set of agents with their attributes and behaviors; (2) a set of relationship and methods of interaction between agents; and (3) the environment. Briefly in ABM, agents may be any entity that pursues a certain goal, act independently of each other and pursue their own objective. But in the presence of different characteristics and interactions, agents will use adaptive behavior where they adjust their behavior to the current status of themselves, of other agents and of their environment. Furthermore, in other words, ABM focuses on modeling behavior of agents and, at the same time, observing and understanding the behavior of the system made up by the agent.

3 The Conceptual Model for Milk Supply Chain Risk

3.1 The Concept of Supply Chain Risk

To define the concept of supply chain risk management, Jüttner [16] and Juttner et al. [13] presented a theoretical framework as shown in Fig. 3. In this framework, it distinguished four basic constructs for supply chain risk management: (i) supply chain risk sources (or disruptive events), (ii) supply chain structure, (iii) risk-mitigating strategies (management), and (iv) risk consequences (outcome). In fact, the level of impact and the consequences of each supply chain are the result of supply chain structure, the magnitude and profile of disruptive event (or risk source), and also the coping and mitigation strategies that are in place.

Based on this theoretical framework, the conceptual model for supply chain risk should be developed in following steps: (i) supply chain modeling, to define

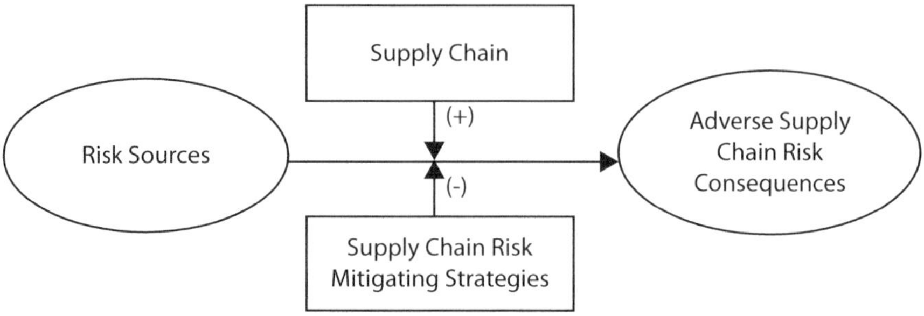

Fig. 3 Supply chain risk analysis framework (Adapted from Juttner et al. [13])

the structure and behavior of a supply chain; (ii) risk modeling, to describe the characteristics of risk or disruptive events; and (iii) risk management modeling, to predict how management practices can be modeled. Among three components above, the supply chain model conceptualization is central, and other two modeling components are constructed based on the former model.

3.2 *The Effect of Risks for Agents in Milk Supply Chain*

The model of risk management in a particular supply chain system will largely depend on agents' behavior. North and Macal [15] illustrated that an agent is basically an individual who has a set of attributes and behavioral characteristics.

In the authors' term, the attributes define what an agent is, and the behavioral characteristics define what an agent does. Thus, an agent can be described by its state and behavioral rules. In fact, the existence of attributes and characteristics will allow agents to take in information, process the inputs, and then affect changes in a particular system and also changes in outside environment. The illustration of this dynamic process can be depicted in Fig. 4.

In this context, risks can be analogous to the inflow of information (in any form) from the outside "world" to the inside of agent. Basically, agents will always monitor the dynamics of external situation and also the current state the agents have. Based on the available information, from outside world or inside (attribute), agents decide what to do and subsequently take one action. The information the agents have, and the taken action, actually are kept in the agent's memory to be deployed under the subsequent updated state. The action that the agents' take, in turn, will become triggers for the system reaching into a new state of system.

However, so far, it is not yet clear how the agents will perceive and process the risks with the above similar arrangement. As previously discussed, Juttner [16] has provided an analytical framework which presents the relationship between the sources of risk and the consequences of any given risk, with the structure of supply chain and it mitigating strategies as determining factor. Yet, how the risks will pass

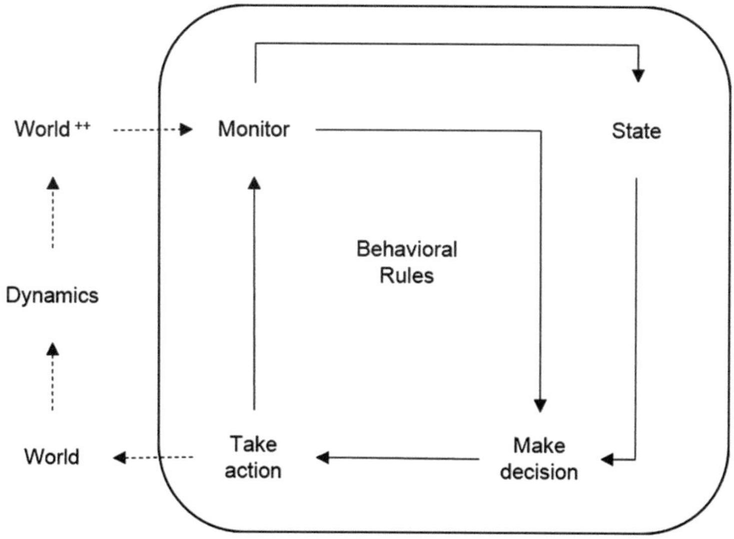

Fig. 4 Generic structure of agent (Modified from Behdani et al. [14]

the effect to every actor in the chain is still indistinct. In the term of agent-based model, these risks affect the state and the behavior of the agents, which in turn, will affect the whole chain system. As the system change, new system's outcome emerges, as in CAS.

For this purpose, additional attribute/properties that are related to risk terminology should be introduced, which are the expected loss and the vulnerability. [17]) defined the expected losses as "a function of the probability of a risky event actually occurring and the exposure to that risky event". A risky event is an event which actually occurs that may influence the outcome. Then, expected losses describe the potential severity of negative impacts from a given event. These losses can be in the form of tangible and intangible, and also in the term of interval, short and long term losses.

In the terminology of agent, the expected losses represent the subject that an agent considers before it makes a decision making to take an action. Just like an information, the agent monitors the probability of any given event that is perceived as risky. Also, the agent will process the information along with the current state and the memory of the agents. In the agents' expected loss, the related main concept is severity.

Severity is a subjective concept which refers to the state of being severe. In fact, an agent, or an actor in the real world, has very limited information about the likelihood of any negative outcomes in the future event; thus, it is very subjective. However, for this purpose, a list of possible outcomes for the expected loss can be derived. One of these outcomes should represent the current state of an agent in the modeling process. Figure 5 illustrates the outcome of expected loss.

		Potential of negative impact / severity	
		Low	High
Probability of risky event	Low	Low probability, low impact	Low probability, high impact
	High	High probability, low impact	High probability, high impact

Fig. 5 Possible outcome of expected loss attribute (Adapted from Jaffe et al. [17])

		Capacity to manage risk	
		Low	High
Expected loss	Low	Low vulnerability	Very low vulnera-bility
	High	High vulnerability	Low vulnerability

Fig. 6 Possible outcome of vulnerable attribute (Adapted from Jaffe et al. [17])

Four possible states for expected losses can be identified, varying from high impact and probability to low impact and probability. Especially in agricultural production, farm assets and their allocation will be determined by the exposure of farms to risk. Jaffe et al. [17] showed that assets allocation which involved in mixed farming (crop and livestock), or diversification of farm and non-farm activities influence exposure to risk, in turn, influenced by risks. Then, both asset allocations and exposure to risk determine the degree of severity in particular risky event.

In conjunction with the concept of expected losses, vulnerability is another concept that should be able to represent the agents' behavior. In this context, OECD [18] defined vulnerability as a function of expected loss and management capacity. For showing direct relationship between the two concepts, Fig. 6 lists the characteristics of vulnerability.

As the concept of expected loss, vulnerability can be a part of an agent's state, which in turn will determine the next action that an agent takes (see Fig. 7). However, the component of management capacity within vulnerability indeed has broader dimension. As accompanied by [17]), they stated that the vulnerability of individual chain participants and the overall supply chain depends on the nature of the risks and on the effectiveness of the risk management instruments in use. Therefore, risk combined with the risk management responses leads to performance outcomes. In this situation, risk management refers to the instruments which could be social or physical instruments that an agent owns or operates. Thus, while expected losses can be a representation of agents' state and behavior individually, vulnerability will be a more collective representation.

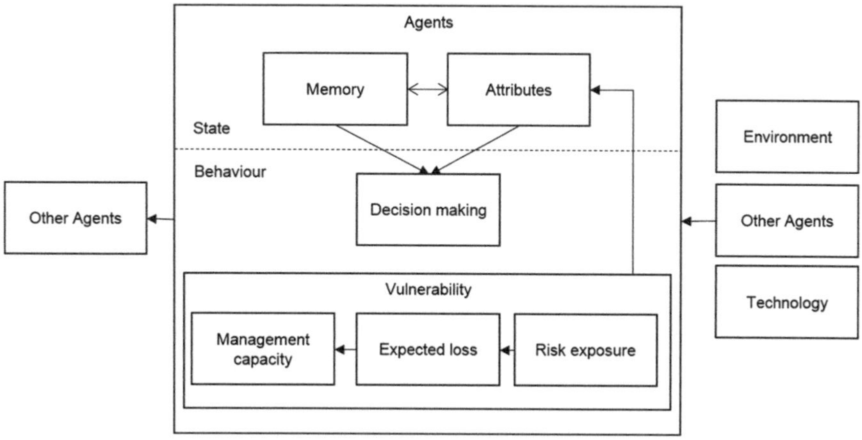

Fig. 7 Structure of agent in the model (Modified from Behdani [14])

3.3 The Basic Model for Milk Supply Chain Risk

The concept of expected loss and vulnerability will now be integrated into the basic model of supply chain to develop a model of supply chain risk management. This integration can be done by modifying the attributes and properties of available agent in order to induce the specific behavior related to the risk and risk management in the system. And in turn, how the agents' behavior will affect the performance of the system is a major importance in this research. The structure of the agent with specific risk-related attributes is presented in Fig. 8. It should be noted that, in this proposed model, farmers and farmers' suppliers will be the agents that deal with risks mostly. This can be justified by actually recognizing farmers as the main producer in the milk supply chain and can be viewed as a push production process (make to stock). However, the other agents, cooperative and manufacturer, will also deal with risks but very specific to organization risks.

Figure 8 presents an experimental design that is proposed in this paper. It also presents the high level of abstraction of decision-making structure that can be modeled. The abstract shows the stage of decision that each agent makes in regard to the flow of products. As can be seen in the figure, the main producers (farmers) will produce raw milk under risks or the perception of risky events. In our previous study, we found several risks (risky events) often faced by farmers [19]. Fodder shortages in dry season and the lack of replacement cattle are the most significant risks perceived by farmers. The poor condition of milk handling and bulking practiced by farmer's cooperative is also perceived as risk that discourages individual milk production.

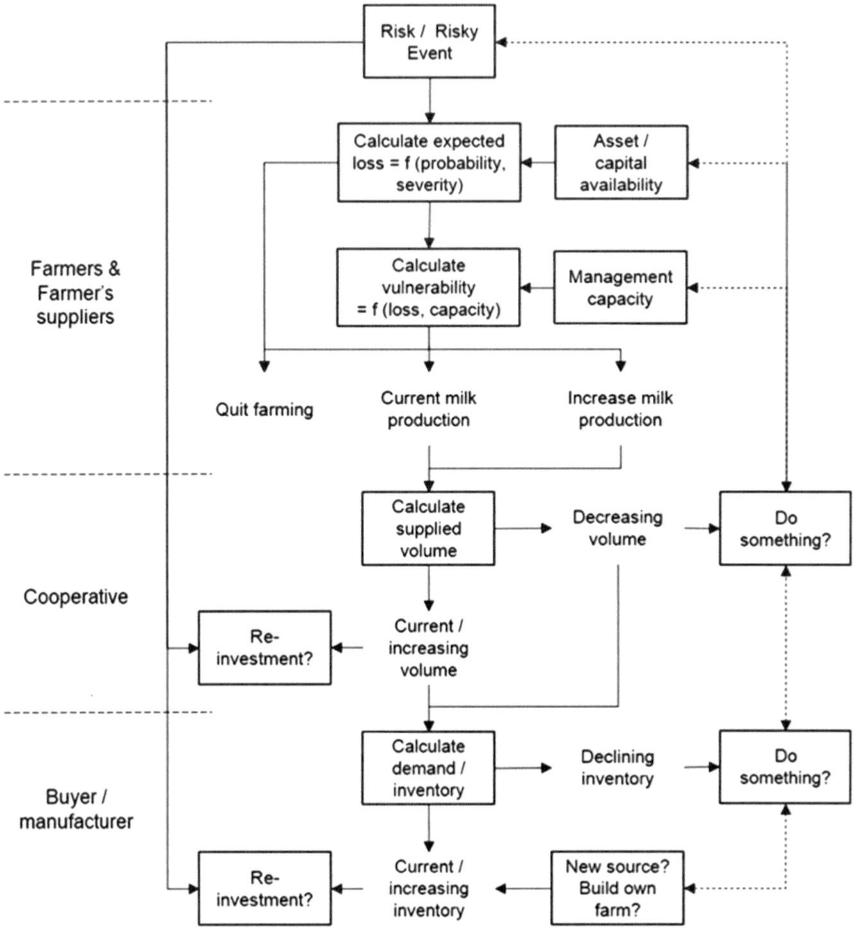

Fig. 8 Agent-based model for milk supply chain risk

In these circumstances, it is believed that there are three possible outcomes for producers, which are (i) staying on the current level of milk production, (ii) somehow increasing the level of production, (iii) or even exiting the dairy business. These outcomes are resulted from the decision-making process that is based on the state of vulnerability consideration. As aforementioned, vulnerability for the main producer is a function of probability of loss, severity, and also risk management capacity. In this case, it should be assumed that if the vulnerability state is relatively low, then the producer will continue to produce raw milk while tending to exit the production if the state is considerably high.

Let's assume that, for any reasons, there are always a small number of producer who quit dairy business over time. For sure, this will lead to gradually decreasing volume of raw milk supplied from the farmers to the cooperatives, which play as milk distributor. This means that the volume that has to be delivered by distributor to the buyer also decreases. As distributor performs a fully logistical function, the volume of products that is forwarded from upstream to the downstream should be the main concern for this enterprise. The consequences from lessening volume of goods will also be experienced financially. To perform day-to-day logistics, it is for sure that a significant level of investment has been realized by the distributor based on the actual and potential volume of products. Thus, if the decreasing level of products supplied from the producers can be assumed to be true, the basic fixed cost to perform such logistical function will be likely increased and weaken the competitiveness of the enterprise. In this setting, this is the main risk which has to be confronted by the cooperative.

The situation is also true for the buyer, which in this case is the manufacturer. This enterprise is the one that confronts the demand for final good from the market. To fulfill its market demand, this enterprise will largely depend on the level of inventory, which is the raw milk received from the distributor. Especially in food industries setting, the raw material (agroproducts) is the most significant component in the production line, both the quantity and quality. As this enterprise sources almost all of its raw material from particular distributor, the decreasing level of raw supply from the distributor will likely also generate undesirable consequences. For handling the customer's orders, the manufacturer will make sure that inventory level is adequate. In this case, gradually decreasing level of inventory will become a significant risk.

As in abovementioned, the proposed model intends to capture and explain how the risks in milk supply chain manifest itself and are transferred throughout the supply chain. Also, this model is expected to present the clear relationship between risks or risky events and the production behavior of each agent in the model. Therefore, the variables that describe the goal of the model are mainly the quantity and the quality of raw milk produced by farmers as main producers and the cost of raw milk collection and distribution incurred by the cooperative and of processing raw milk into dairy products by the manufacturer. Essentially, since changes of flow of raw milk will have a great possibility for disrupting the whole production process and lessen the financial performance of every actor in a particular supply chain, then these variables are expected to provide main indicators for the dynamics of the whole supply chain.

However, the focus of the model is also to find the emergence in the production system given the presence of risk. The "do something" box in Fig. 8 actually represents if there are any efforts from the downstream actors (in this case are the distributor and manufacturer, respectively) to collectively deal with the risks that expose upstream actors. If the previous hypothetic condition is assumed true, it can be seen that decreasing level of production in the upstream actors caused by risks will become systemic risk, or supply chain risk, because it manifested throughout the chain. In this case, the manufacturer has to do something to secure its raw materials supplied from the distributor, and in turn, the distributor does the similar on the main producer to guarantee the capability of its logistical function. Similarly, the main producers have to keep the state of vulnerability low in order to continue the production. For example, to overcome fodder scarcity faced by farmers, the downstream actors could involve in the provision of commercial fodder and other input as well. Therefore, risky events associated with the uncertainty of input availability can probably be reduced in the upstream level. Thus, it is one of the main concerns of this research to present these phenomena through the proposed model.

4 Summary

A basic agent-based model for assessing supply chain risk has been introduced in this paper, specifically in milk production. The agent-based approach used in conceptualizing the model is proposed given that its flexibility to accommodate the real phenomenon is very complex. In the real world, milk supply chain has many stages of production performed by various actors with the different levels of technology and in the different time, thus consisting of complex decision-making structure. The model can conceptually demonstrate that risks faced by upstream actors will likely be transferred to downstream actors throughout the chain through the level of production as the outcome of given risk management practiced by each actor. It is possible that the production outcome of one actor in one stage becomes the source of risk for later actors in the later stages. This condition implies that risk management in a particular supply chain should be practiced in a systemic manner involving all actors in the milk supply chain.

References

1. Hardaker, J., Huirne, R., & Lien, G. (2004). *Coping with risk in agriculture*. New York: CAB International.
2. Holton, G. A. (2004). Defining risk. *Financial Analysts Journal, 60*(6), 19–25.
3. Bachev, H. (2012). Risk management in the agri-food sector. *Contemporary Economics, 7*(1), 45–62.
4. Van der Vorst, J. G. A. J., da Silva, C. A., & Trienekens, J. H. (2007). *Agro-industrial supply chain management: Concept and application*. Rome: Agricultural Management, Marketing and Finance. Food and Agriculture Organization (FAO) of the UN.
5. Railsback, S. F., & Grimm, V. (2012). *Agent based and individual based modelling*. New Jersey: Princeton University Press.
6. Bebe, B. O., Udo, H. M. J., Rowlands, G. J., & Thorpe, W. (2003). Smallholder dairy systems in the Kenya highlands: Breed preferences and breeding practices. *Livestock Production Science, 82*, 211–221.
7. Bryceson, K, & Slaughter, G. (2009, March 25–27). Integrated autonomy – A modeling-based investigation of agrifood supply chain performance. Proceeding of 11th international conference on computer Modelling and simulation (334–339). IEEE Computer Society, Cambridge, UK.
8. Mishra, P. K., & Shekhar, B. R. (2011). Impact of risks and uncertainties on supply chain: A dairy industry perspective. *Journal of Management Research, 3*(2), 1–18.

9. Choi, T. Y., Dooley, K. J., & Rungtusanatham, M. (2001). Supply networks and complex adaptive systems: Control versus emergence. *Journal of Operations Management, 19*(3), 351–366.
10. Pathak, S. D., Day, J. M., Nair, A., Sawaya, W. J., & Kristal, M. M. (2007). Complexity and adaptivity in supply networks: Building supply network theory using a complex adaptive system perspective. *Decision Sciences, 38*(4), 547–580.
11. Surana, A., Kumara, S., Greaves, M., & Raghavan, U. N. (2005). Supply chain networks: A complex adaptive system perspective. *International Journal of Production Research, 43*(20), 4235–4265.
12. Lazzarini, S. G., Chaddad, F. R., & Cook, M. L. (2001). Integrating supply chain and network analyses: The study of netchain. *Journal on Chain and Network Science, 1*(1), 7–22 . 12.
13. Juttner, U., Peck, H., & Christopher, M. (2003). Supply chain risk management: Outlining an agenda for future research. *International Journal of Logistics: Research and Applications, 6*(4), 197–210.
14. Behdani, B. (2013). Handling disruptions in supply chains: An integrated framework and an agent-based model. Ph.D thesis, NGInfra PhD thesis series on infrastructures no 54, Delft University of Technology, Netherlands.
15. Macal, C. M., & North, M. J. (2010). Tutorial on agent-based modeling and simulation. *Journal of Simulation, 4*, 151–162.
16. Jüttner, U. (2005). Supply chain risk management: Understanding the business requirements from a practitioner perspective. *The International Journal of Logistics Management, 161*, 120–141.
17. Jaffe, S., Siegel, P., & Andrews, C. (2010). Rapid agricultural supply chain risk assessment: A conceptual framework. Agriculture and rural development discussion paper no 47 The World Bank.
18. OECD. (2009). *Managing risk in agriculture; a holistic approach*. Paris: The Organisation for Economic Co-operation and Development (OECD) Publishing.
19. Daud, A. R., Putro, U. S., & Basri, M. H. (2015). Risks in milk supply chain; a preliminary analysis on smallholder dairy production. *Livestock Research for Rural Development, 27*, Article #137.

People's Willingness to Donate Blood: Agent-Based Approach

Dania Nurissa Dwiartika, Utomo Sarjono Putro, Manahan Siallagan, and B. S. Onggo

Abstract In cases where a trial-and-error experiment is costly or impossible, especially in healthcare industry, researchers have used simulation modeling to avoid the risk caused by a trial-and-error experiment. In healthcare industry, blood supply plays an important role because the shortage of blood could make people's life at risk. In most countries in Southeast Asia, including Indonesia, blood services have not been considered as an essential service for healthcare support program. Moreover, blood supply chain in low-and-middle-income countries has different characteristics and challenges compared to the high-income countries. For developing countries, one of the important factors is the number of donors. This research conducted to see how the agent makes decision about donating their blood. Finding?

Keywords Agent-based • Theory of planned behavior • Willingness to donate • Healthcare • Blood donor

1 Introduction

There are several cases in healthcare where a trial-and-error experiment is impossible to be carried out. For example, a trial-and-error experiment is unethical when it involves someone's life. Nowadays, proper healthcare or medical facilities are essential facility that should be managed responsibly and usually is monitored

---

D.N. Dwiartika (✉)
Bandung Institute of Technology, Bandung, Indonesia
e-mail: dania.nurissa@sbm-itb.ac.id

U.S. Putro • M. Siallagan
School of Business and Management, Institut Teknologi Bandung, Bandung, Indonesia
e-mail: utomo@sbm-itb.ac.id; manahan@sbm-itb.ac.id

B.S. Onggo
Lancaster University, Lancaster, UK
e-mail: s.onggo@lancaster.co.uk

U.S. Putro et al. (eds.), *Agent-Based Approaches in Economics and Social Complex Systems IX*, Agent-Based Social Systems 15, DOI 10.1007/978-981-10-3662-0_2

closely by the authority [1]. Therefore, the demand for research in healthcare modeling is increasing. Many researchers have used simulation modeling to solve issues in healthcare. Researchers believe that simulation modeling is an interesting decision support technique that could be very helpful to the decision maker. Although simulation has been widely used for decades in another area of industry and manufacturing, it is relatively new for applying simulation in healthcare context.

In the healthcare context, blood is the most vital perishable product used [2]. The characteristics and problems in the blood supply chain faced by the developing countries are different with the developed country [3]. A common problem in blood supply chain in a low-and-middle-income country (LMIC) is the availability of blood supply. A shortage in blood supply can put life at risks. Despite all of that, in many developing countries such as Indonesia, blood services have not been considered as an essential service for healthcare support program and the research about this blood supply chain is rarely done.

The availability of blood in blood centers is very important. The blood will be used for transfusion when needed (mostly urgently), e.g., accidents, casualty, major surgery, and thalassemia patient. In fact, through prior research investigation, we know that PMI (Palang Merah Indonesia), Indonesian Red Cross organization, has experienced shortages of blood supply. Among 211 PMIs in 2010, they could only meet 60% of the blood needed. Moreover, the blood collected by 140 hospital-based blood centers is far from meeting the expectation; not all the hospitals are functioning as it should be [4]. By functioning, it means that they do not have the standard quantity. Therefore, for developing countries, one of the important factors is the number of donors.

This objective of this exploratory research is to understand people's willingness to donate blood and to find out factors that affect people's decision to donate blood. We want to get insights and a fuller understanding about the people's willingness to donate blood in the context of a developing country. Samples are collected from Bandung in Indonesia. The data will be used to calibrate our agent-based simulation model. The agent-based simulation is used to explore how the interactions among donors and potential donors affect the supply of blood.

2 Literature Review

2.1 *Willingness to Donate*

For predicting the number of people who donate their blood, there are two factors that influenced them, intention and past behavior. Intention is the best predictor for novice donors (including the first donor and donors who have been donating their blood less than four times). Theory of planned behavior formulated a person's intention from three factors: (1) beliefs, how they form beliefs about the consequences of an action; (2) subjective norms, how they perceived social pressure; and (3) perceived behavioral control, how the person perceived their ability to take the action [5]. Past behavior is a significant predictor for a regular donor (more than four time donations) [6]. Other researchers argue that a direct measure of past behavior improves behavioral predictions of blood donation [5]. Another framework that is widely used in social science and public health context is the theory of reasoned behavior [7]. The theory states that intention is a combination of social norms and attitude [8].

Other factors that influenced people's willingness to donate are their motivation, either health motivation or benefit-seeking motivation or altruistic motivation [9]. By the term motivation, we mean the process of activating the individual in order to reach a goal (objective), taking into consideration the conditions of the environment in which she/he is situated [10]. From some prior research, motivation is one of the factors that drive people's willingness to donate [10, 11]. For doing something voluntarily, there are some motivation which includes social motivation, value motivation, self enhancement motivation, ego-protection motivation, and knowledge motivation [10]. Several researchers state that the use of communication factor, such as word of mouth, might be helpful to push people to donate their blood [9, 12, 13]. Observing other behaviors (observer effect) has the same impact with the word of mouth effect to people's willingness to donate. Observer effect can negatively affect people's willingness to donate, if they observe many of negative effects happen after the blood donor procedure, e.g., fainting [5].

2.2 Simulation in Healthcare Industry

As it is said in the introduction, even though simulation has already been widely used in various fields, military, manufacture, and telecommunication, the simulation approach is relatively new in the healthcare industry. Here are the previous studies, using simulation and modeling, that studied major themes in the healthcare especially blood supply chain.

Discrete-event modeling is a common method used by the researcher in the healthcare industry. It is because of discrete-event modeling approach based on the concept of entities, resources, and describing entity flow [14]. Discrete event is also commonly used for modeling the blood supply chain for the supply chain efficiency [1, 2, 15]. The use of discrete event simulation resulted in an optimal inventory control and removing overlapping functions [2].

Another approach is a system dynamics approach. System dynamics is one of the primary approaches in the simulation and modeling topic. SD abstracts based from a single event and takes the holistic point of view to observe the policy [16] and based on the philosophy that a system's structure leads to systems behavior [17]. Samuel et al. [17] said that system dynamics could portray the accumulation and feedback which can be tested systematically to find a possible scenario for developing policies [18]. One of the examples using the system dynamics in the blood supply chain is trying to investigate the performance of the blood supply chain system itself [19]. System dynamic focuses on how to understand the nature of the whole system behavior, not to explore the effect of uncertainty [25].

There are several other methods that have been used for blood supply chain modeling and simulation, network optimization model [20], geographic information system (GIS) [21], game-based approach [22], etc. For researches that focused on people's willingness to donate blood, the researcher mostly used the statistical modeling, a modeling approach that focuses on how a system's outputs depend on its inputs, without regard to the system's internal structure or causal processes [25].

This research used the simulation experiment – agent-based modeling. We use the agent-based modeling because it can define behavior at the individual level and the global behavior emerges as a result of many (tens, hundreds, thousands, millions) individuals, each following its own behavior rules, living together in some environment and communicating with each other and with the environment [14]. As mentioned above, one of the problems in developing country is how to maintain the number of donors because the number of donors will be affecting the number of blood bags in the blood center, whereas the blood donors have various behaviors. These variations in the donor's characteristics and behaviors made a simulation using standard variables not enough. The agent perspective is important to be taken, so using agent-based modeling will be effective in that way. An agent-based model shows how a system could evolve through time in ways that are often difficult to predict from knowledge of the behaviors of the individual agents alone [25].

3 Modeling

In blood supply chain, there are three stakeholders that interact with each other. There are hospital, blood center, and the donor itself. The interaction among each stakeholder is not hierarchical because the interaction could be intertwined and so on (Fig. 1).

Fig. 1 Modules interaction

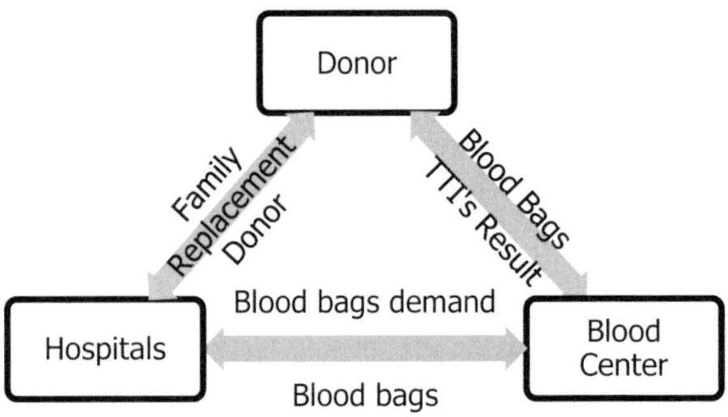

3.1 Model Building

This paper only focuses on the donor module. For the donor module we used the simulation experiment – agent-based modeling. We use the agent-based modeling because it can define behavior at the individual level, and the global behavior emerges as a result of many (tens, hundreds, thousands, millions) individuals, each following its own behavior rules, living together in some environment and communicating with each other and with the environment [14]. In this model, the main concern is to describe how the uncertainty of the person's behavior could be affecting the number of donors. With some intervention in the model, we could see how to encourage people to become a donor. In this module, the main concerns are how to encourage people to become regular donors and how to keep the current repeated donors [3]. Agent-based modeling is chosen because we want to observe the number of donors so the output from this module could be stated as the number of blood bags that a blood center could get, if we assumed that one person only donate one blood bag each time they donor (Fig. 2).

This framework illustrates core rule in the model simulation. Every agent owns their own condition, this including their belief and health condition. From their own condition, they can have their willingness to donate, this willingness shown by the intentional behavior. Whether the agent donating or not donating their blood, it will be affecting the number of people who have been donating their blood. Furthermore, the agent will be observing the behavior of people who have donated their blood. If the observed people have a negative effect after the donation, it will be negatively affecting the agent. Other than that, the decision will be affecting their own condition, like health condition and belief.

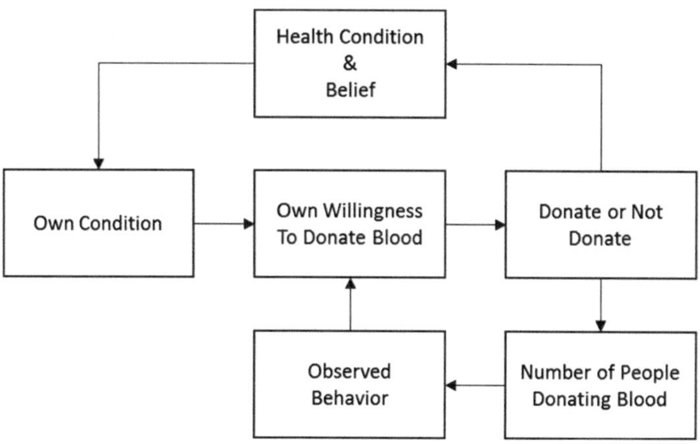

Fig. 2 Model framework

There is a set of assumptions that will be used in this model.

1. Agents. Agent in this simulation will be divided into non-donor, potential donor, and actual donor. Potential donors are the people who have the intention to donate blood, but they have not actually donated their blood. There are several causes such as their health condition and the location of the blood center is far away. Actual donors are the agents who actually donate their blood. The number of blood bags a blood center could get will be counted from the number of actual donors.
2. 2% of all the donors will have a negative effect after donating their blood/vasovagal reaction (VVR) [6]. From other researches, the percentage of a new donor which will have a vasovagal reaction is between 2% and 9% [6, 23].
3. Every time the agents donate their blood, their health condition will improve. Medical Post states that several health researchers mention blood donors as one action that could lower the level of iron in body, so eventually it could lead to decreasing the cardiovascular risk.

There are several attributes for the agents. This attribute shows the characteristics of each agent (Fig. 3).

1. Health condition (HC)
   Every donors/potential donor health condition will affect whether they can donate their blood or not. This factor varies from 0 to 1.
2. Belief (B)
   The value varies from 0 to 1. The higher it is means that they have strong beliefs about the act of blood donation.
3. Subjective norms (SN)
   This factor represented by social pressure, the value varies from 0 to 1. If more people donated their blood, the values will likely approaching 1.
4. Perceived behavioral control (PBC)
   This is how the agent perceived their own condition. Whether they could make the donation or not. The fear of needles or the fear of blood most likely be the driven factors affecting this attribute. The value varies from 0 to 1; the higher the value means they could easily donate their blood.
5. Intentional behavior (IB)
   This factor portrays people's attitude toward blood donor. This factor affected from belief, perceived behavioral control, and subjective norms. The higher the intention it is more likely for people to become an adoptee (in this case became a potential donor).
6. Mobility (M)
   This factor represents the people willingness to go to a blood center. The lower the value means that the people felt burdened to go to a blood center.

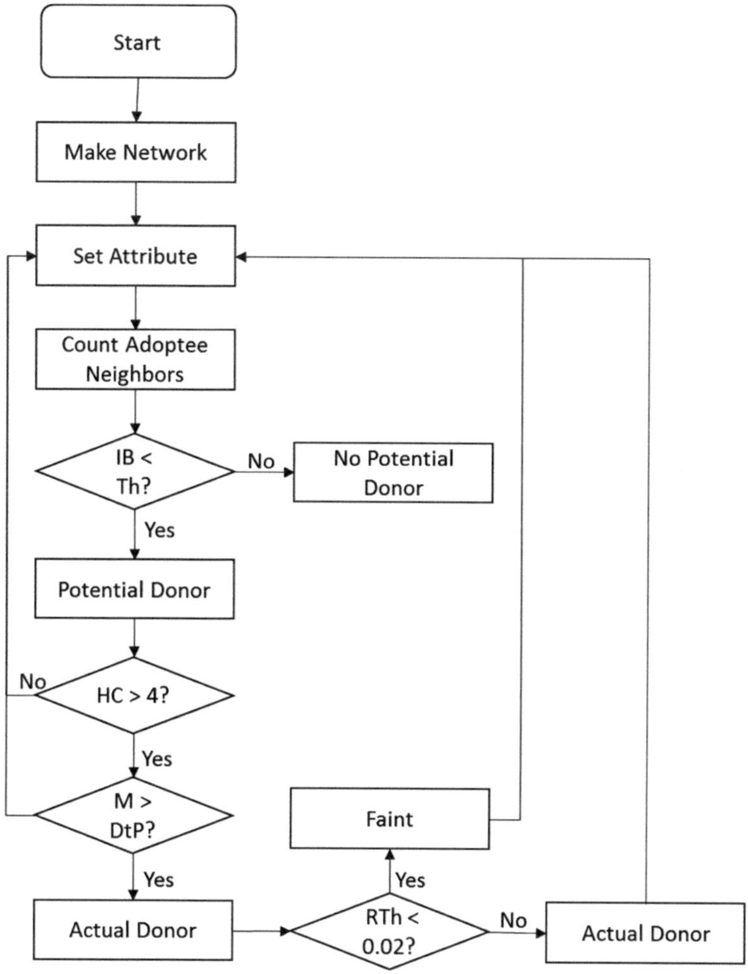

Fig. 3 Simulation flowchart

3.1.1 Decisions of Agents

From the theory of planned behavior and the theory of reasoned behavior, intentional behavior will be the best predictor to the actual behavior. This intentional behavior depends on the agent's belief and subjective norm. Belief in the blood donation is often associated with the attitude toward blood donation. The stronger the belief, the more favorable the attitude toward the act of blood donation. That is why the intentional behavior will be used as a threshold for people to become a potential donor (adoptee).

$$IB = B \times SN \tag{1}$$

and

$$SN = 1 - \exp\left(\frac{\text{adoptee neighbors}}{\text{total neighbors}}\right) \tag{2}$$

Using the diffusion theory, this intentional behavior will become a threshold. If the threshold is smaller than the intentional behavior, the agent will become an adoptee for blood donor action, or we called it potential donor and vice versa. Below is the equation for determining the agent as a potential donor.

$$IB < \text{Random Threshold} \tag{3}$$

and

$$IB \geq 0 \tag{4}$$

Subsequently a potential donor (adoptee) will become a donor if their health condition is sufficed and the distance to blood center (DtP) is lower than their own mobility. Another effect that influences their donation is the awareness of their own health condition. For every time they donate, their health condition will increase. As stated from the PMI, the blood donation will make the process of renewal of the red blood cell faster, increasing the immunity system in the donor body.

$$HC \geq 4 \tag{5}$$

$$M > DtP \tag{6}$$

For every donor, 2% of the total donor will experience the vasovagal reaction (RTh = Random Threshold). This negative effect will have an impact for the observer effect. The number of other donors observed to faint was significantly predicted future behavior of donor [6]. This number of donors negatively affects the future donation.

$$HC_{(t+1)} = 0 \text{ if } RTh \leq 0.02 \tag{7}$$

$$HC_{(t+1)} = HC_{(t)} + 0.01 \text{ if } RTh > 0.02 \tag{8}$$

An agent who has a zero health condition will turn back into a non-donor status, and it will be influencing the subjective norms. This update on agent's health condition will eventually affect the belief which influences the intentional behavior.

$$B_{(t+1)} = B_{(t)} + \alpha\left(R_{(t)} - B_{(t)}\right) \tag{9}$$

α is a parameter which varies between 0 and 1 and will be adapted from the actual condition. $R_{(t)}$ is rewarded for every agent for each iteration; in this simulation, the reward is set to 1. If the agent feels satisfied from the reward they get, the probability of the intentional behavior will be bigger than before and vice versa.

4 Result and Analysis

The result from the simulation is showed below. In this simulation, the number of agents is 100, and the simulation runs for 72 iterations. This iteration represents a blood center that runs for 6 years (72 months). This time period is chosen to match the actual data that is available (2009–2014). The result shows that the number of donor fluctuates with the highest number of 37. We will validate the model with this output and compare the data with the annual data from PMI (Figs. 4 and 5).

The model validation is using the F test. $H_0 : \sigma_1^2 = \sigma_2^2$ and $H_1 : \sigma_1^2 \neq \sigma_2^2$, if the p-value is significant it means we reject the H_0. The F test is conducted using the data analysis in excel. It is shown in Table 1. In this table, we could see that the p-value is much smaller than 0.05. It means the data is significant. We can conclude that we reject the H_0 which means that the variance between the two data is different.

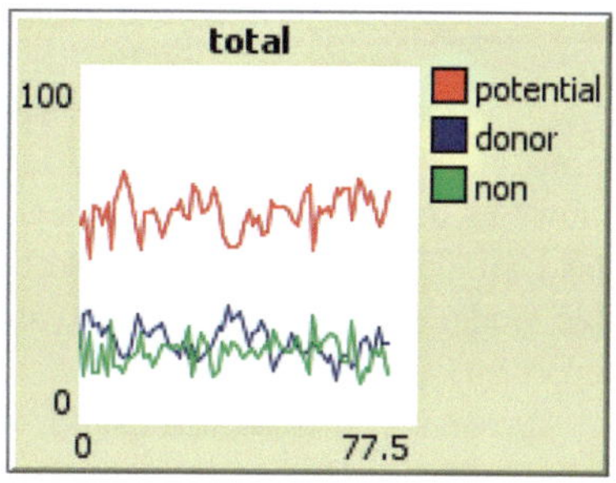

Fig. 4 People's willingness to donate

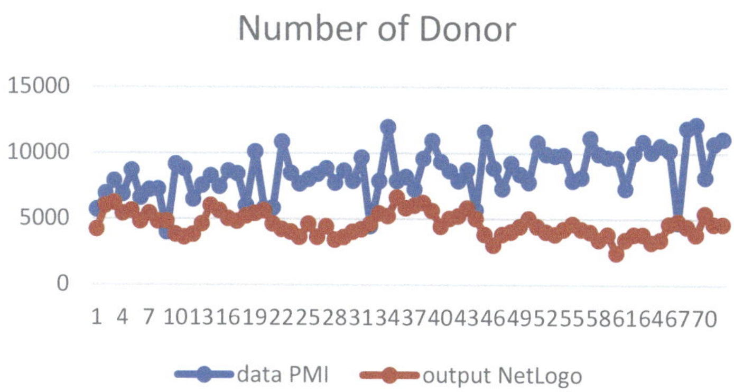

Fig. 5 Model validation

Table 1 F-test two samples (simulation output and PMI data) for variances

	Variable 1	Variable 2
Mean	8534,583333	4558,333333
Variance	3271796,894	768380,2817
Observations	72	72
df	71	71
F	4,2580438	
P(F<=f) one-tail	2,53575E-09	
F Critical one-tail	1,481482322	

Table 2 t-test: two-sample assuming unequal variances

	Variable 1	Variable 2
Mean	8534,583333	4558,333333
Variance	3271796,894	768380,2817
Observations	72	72
Hypothesized mean difference	0	
df	103	
t Stat	16,78571035	
P(T<=t) one-tail	1,52555E-31	
t Critical one-tail	1,659782273	
P(T<=t) two-tail	3,05109E-31	
t Critical two-tail	1,983264145	

Table 2 showed the result for t-test. This t-test is to test if there are any differences in mean or not. $H_0 : \mu_1 = \mu_2$ and $H_1 : \mu_1 \neq \mu_2$. From the table above, the p-value shows us that we reject H0. It means the mean between the simulation output and the PMI data is different. Those two results of the F-test and t-test showed that this model cannot be validated. Validation itself means making sure that the model represents the real world.

Apparently this model cannot support the previous model. Even though the behavior is built based on some theory, the implementation of how the theory could be changed into the biased behavior in agent should be rethought. This model just uses the theory as it is, i.e., belief defined as belief. It could be that there are other factors that built the agent belief. Godin et al. [25] said that belief or attitude is based on the cost-benefit of the act such as moral responsibility, personal obligation, etc. Another factor could be the external interaction. As it can be seen in the model, there is no interaction among the agents, unless just observing others' behavior. Scherer and Cho [24] said, "Individual adopts the attitudes or behaviors of others in the social network with whom they communicate." Apparently people will likely become a potential donor if the people around them share the same attitudes, information, and beliefs. Zhou et al. [12] state that the advertisement can be used to increase the motivation to donate blood, particularly when the advertisement appeals match the person's motivation.

5 Conclusion

Blood services are not considered as an essential service for healthcare support program in many low-income countries including Indonesia. Regarding the importance of blood service such as maintaining the availability of blood stock, calculating or predicting the number of donors each month is something that is vital for low-income countries. The number of donors could be observed through the behavior.

Human decision making is a complex process. This has also been observed in our study on people's willingness to donate blood. Based on theory of planned behavior and theory of reasoned behavior, we can explain that the willingness to donate blood is based on the person's intention [5, 8]. This intention is affected by the person's belief/attitude, subjective norms, and how they perceived their capability of giving blood. This paper has found that peoples' willingness to donate blood in our study is affected by belief, subjective norms, perceived behavioral control, health condition, and mobility. We have also developed an agent-based model to study how the interactions between donors and potential donors affect the supply of blood. The model has produced some interesting outcomes. What are the outcomes?

This study has a number of limitations that we will address in our future works. First, we only assign a simple trait to the agents. A more complex trait might be used to better represent the behavior of donors and potential donors such as belief or attitude that is based on the moral responsibility and personal obligation [25]. Secondly, the agent is assumed to have a willingness to donate blood based simply on their intentional behavior. In the real life, there are other factors that make people want to donate blood such as advertising and anticipated regret (negative feeling toward the idea of not giving blood). Thirdly, the agents are assumed to interact by observing others' behavior, through the lens of a subjective norm. In the real life, individuals could meet with a donor and their belief/attitude about the act of donation may change. Finally, the validation is a challenge. Hence, the model in this paper should not be used as a forecasting tool (i.e., to estimate the number of donors), but it should be used as a tool to explore how different behaviors may impact the supply of blood.

References

1. Rytila, J. S., & Spens, K. M. (2006). Using simulation to increase efficiency in blood supply chains. *Management Research News, 29*(12), 801–819.
2. Yuzgec, E., Han, Y., Nagarur, N. (2013). A simulation model for blood supply chain systems. *Proceedings of The 2013 industrial and systems engineering research conference.*
3. Onggo, B. S. (2014). Elements of a hybrid simulation model: A case study of the blood supply chain in low-and middle-income countries. *Proceedings of the 2014 winter simulation conference.*
4. Soedarmono, Y. S. M. (2010). Donor issue in Indonesia: A developing country in South East Asia. *Biologicals, 380*, 43–46.

5. Holdershaw, J., & Gendall, P. (2011). Predicting blood donation behaviour: Further application of the theory of planned bahaviour. *Journal of Social Marketing, 1*, 120–132.
6. Ferguson, E., & Bibby, P. A. (2002). Predicting future blood donor returns: Past behavior, intentions, and observer effects. *Health Psychology, 21*(5), 513–518.
7. Davis, C. A., et al. (2015). A role for network science in social norms intervention. *Procedia Computer Science, 51*, 2217–2226.
8. Fishbein, M., & Ajzen, I. (2010). *Predictiong and changing behavior: The reasoned action approach*. New York: Psychology Press.
9. Tscheulin, D. K., & Lindenmeier, J. (2005). The willingness to donate blood: An empirical analysis of socio-demographic and motivation-related determinants. *Health Services Management Research, 18*(3), 165.
10. Guiddi, P., et al. (2015). New donors, loyal donors, and regular donors: Which motivations sustain blood donation? *Transfusion and Apheresis Science, 52*, 339.
11. Karacan, E., et al. (2013). Blood donors and factors impacting the blood donation decision: Motives for donating blood in Turkish sample. *Transfusion and Apheresis Science, 49*, 468–473.
12. Zhou, Y., Poon, P., & Yu, C. (2012). Segmenting blood donors in developing countries. *Marketing Intelligence & Planning, 30*(5), 535–552.
13. Burdit, C., et al. (2009). Motivation for blood donation among African Americans: developing measures for stage of change, decisional balance, and self-efficacy construct. *Journal of Behavioral Medicine, 32*, 429–442.
14. Borshchev, A., & Filippov, A. (2004, July 25–29). From system dynamics and discrete event to practical agent based modeling: Reasons, techniques, tools. *The 22nd international conference of the system dynamics society*. Oxford, England.
15. Suwardie, A. W., Sopha, B. M., & Herliansyah, M. K. (2013). A simulation model of blood supply chain at Indonesian regional red-cross. *Proceedings of The 2013 international conference on logistics and maritime systems.*
16. Chahal, K., & Eldabi, T. (2011). Hybrid simulation and modes of governance in UK healthcare. *Transforming Government: People, Process and Policy, 5*(2), 143–154.
17. Samuel, C., et al. (2010). Supply chain dynamics in healthcare services. *International Journal of Health Care Quality Assurance, 23*(7), 631–642.
18. Young, M. (1989). *The technical writer's handbook*. Mill Valley: University Science.
19. Afshar, J., et al. (2014). System dynamic analysis of a blood supply chain system. *Applied Mechanics and Materials, 510*, 150–155.
20. Rusman, M., & Rapi, A. (2014). Blood banks location model for blood distribution planning in Makassar city. *Proceedings of the Asia Pacific industrial engineering & management systems conference 2014.*
21. Delen, D., et al. (2011). Better management of blood supply-chain with GIS-based analytics. *Annals of Operations Research, 185*, 181–193.
22. Katsaliaki, K., Mustafee, N., & Kumar, S. (2014). A game-based approach towards facilitating decision making for perishable products: An example of blood supply chain. *Expert Systems with Application, 41*, 4043–4059.
23. Gillet, P., et al. (2015). First-time whole blood donation: A critical step for donor safety and retention on first three donations. *Transfusion Clinique et Biologique, 22*, 312.
24. Scherer, C. W., & Cho, H. (2003). A social network contagion theory of risk perception. *Risk Analysis, 23*, 2.
25. Godin, G., et al. (2005). Factors explaining the intention to give blood among the general population. *Vox Sanguinis, 89*, 140–149.

An Agent-Based Simulation Approach for Government Intervention in Reducing High-Risk HPV

Pri Hermawan, Yuliati Komar, Utomo Sarjono Putro, and Soehartati Gondowiardjo

Abstract This research aims to describe the phenomena of cervical cancer spreading in Indonesia limited to cervical cancer spread caused by the human papillomavirus (HPV) and proposed agent based as an alternative of decision support system for the government. This model focused on the high-risk HPV infection through human sexual behavior. Literature studies, observation, and depth interview with the experts of oncology were implemented. The agent-based simulation (ABS) illustrates the social and sexual interactions in which the HPV may spread easily. It also allows the interaction that occurred if government interventions were implemented and shows how much these actions affect the number of HPV infections. The results gave significant findings on the importance of the government intervention to prioritize on promotion and socialization program on sexual health practice in Indonesia, before implementing both the HPV vaccination (as a national program), and the improvement of the medical treatment on HPV+ patients.

Keywords Decision support system • Agent-based simulation • Human papillomavirus • Infection • Sexual behavior

1 Introduction

Cancer mortality in the world is predicted to increase sharply almost doubled over the past two decades into the future. The latest report from the International Agency for Research on Cancer (IARC) states that the global cancer burden would be a

P. Hermawan (✉) • Y. Komar
Institut Teknologi Bandung, School of Business and Management, Bandung, Indonesia
e-mail: prihermawan@sbm-itb.ac.id; yuliati.komar@sbm-itb.ac.id

U.S. Putro
School of Business and Management, Institut Teknologi Bandung, Bandung, Indonesia
e-mail: utomo@sbm-itb.ac.id

S. Gondowiardjo
Radiotherapy Department, Dr. Cipto Mangunkusumo Hospital, Jakarta, Indonesia
e-mail: gondhow@gmail.com

U.S. Putro et al. (eds.), *Agent-Based Approaches in Economics and Social Complex Systems IX*, Agent-Based Social Systems 15, DOI 10.1007/978-981-10-3662-0_3

major cause of death in the 2010s. It was also predicted that by the year of 2030, 27 million new cancer cases will arise and 17 million cases of cancer deaths will occur each year, while as a comparison in the year 2007 new cancer cases and 12 million cases of cancer deaths numbered about eight million. In 1970 it was stated that only 15% of cancer cases occur in poor countries and developing countries. But today more than 50% of cancer cases and two thirds of cancer deaths occur in this region.

Cervical cancer is the third most common malignancy in women worldwide. Experts have predicted that 370,000 cases per year of cervical cancer in the world occurred with approximately 70% of death rate. In the year 2000, approximately 468,000 new cases were diagnosed, of which 80% was in developing countries. The average annual incidence of cervical cancer varies widely by geographic area, with the highest incidences reported in Latin America, sub-Saharan Africa, and South Asia and Southeast Asia [1].

Indonesia, as one of the archipelagos in Southeast Asia, consists of more than 13,677 islands with a population of 237,641,326 people (BPS data of 2010 Indonesian census). Its capital, Jakarta, situated on the most populous island Java, now harbors over 10 million people. According to expert estimations, the incidence rate of cancer in Indonesia was about 100–190 per 100,000 people. Cervical cancer is the most common malignancy among Indonesian women, up to 22.5% of all cancer cases reported in governmental hospitals [2–4].

Cervical cancer concerns a major health problem in Indonesia since most patients present in late stages of the disease. The association of high-risk HPV types (notably 16, 18, 31, and 45) and the aetiology of cervical cancer are now widely accepted, as HPV has been detected in almost all cervical cancers and much less in controls. Both high-risk and low-risk HPVs can cause the growth of abnormal cells, but generally only the high-risk HPVs lead to cancer. One hundred types of HPV have been detected, 40 spread easily by genital contact, and 15 types of them cause cervical cancer and are defined as sexually transmitted high-risk HPVs (http://hpv.emedtv.com/hpv/hpv-and-cervical-cancer-p2.html).

Thus the first issue to explore is that the topology and dynamics of a sexual network depend upon the cultural context of the respective society. Tendency of clustering, mixing behavior, and the variation in the sexual contacts are important determinants of the network and have significant implications in the spread of high-risk HPV epidemic. The second issue is the topology and dynamics of high-risk HPV infection spread/network which have significant implications of the cervical cancer growth and stages which also give impact whether toward the mortality or the recovering rates of the detected cervical cancer patients. The intervention of this mortality and recovering rates of cervical cancer depends on its medical treatments and is stated as government intervention to prevent or to reduce the impact of cervical cancer for Indonesian women.

In the term of government intervention, the government faces many challenges associated with handling of cervical cancer problem. The Ministry of Health of the Republic of Indonesia mentioned that there are four main issues that need to be enhanced in accordance with the cervical cancer, i.e., (1) the social impact of cervical cancer, which is related to the high financing of treatments that can disrupt the economy of the family, the nation, and the state; (2) cancer as a global epidemic; (3) the effectiveness of cervical cancer treatment, related to the need to evaluate the effectiveness and researching the impact; and (4) the efforts to improve cancer prevention, e.g., facilities of cancer diagnosis and treatment, expanding the scope of early detection of cervical cancer, public awareness that cancer can be prevented, and prevention efforts through prevention of risk factors such as HPV by sexual transmission [Health Ministry News, 2013]. The complexity of cervical cancer problems needs to be explored. Due to its complex system of cervical cancer spread, this research was limited to simulate the spread of high-risk HPV, before any symptoms of cancer occurred.

As the limitation of this research, our simulation provides intervention in the point of view of the Indonesian government (the Ministry of Health of the Republic of Indonesia). Based on observation, there are at least three other agents involved in the intervention of reducing the risk of cervical cancer.

The research questions are the following: (1) What mechanism describes the sexual network to have significant implication on the spread of high-risk HPV epidemic? (2) What will happen if interventions (the sexual health promotion, the HPV vaccinations (national program), and the medical treatment for patients with HPV+) occurred or not?

The purpose of this research is to observe the human behavior impact on the speed of HPV infection spread, to examine the number of HPV+ patients, and to describe the effect of three primary interventions mentioned on point 2 above.

2 The Basics

2.1 High-Risk Human Papillomavirus Transmission as the Primary Cause of Cervical Cancer

This research applied the high-risk HPV transmission theory, which is related to the sexual transmission disease (STD). The STDs are acquired during unprotected sex with an infected partner [3, 5, 6]. Biological factors that affect the spread of STDs include a symptomatic nature of STDs that do not produce or mildly produce any symptoms or signs, gender disparities where women suffer more frequent and more serious STD complications than men do, age disparities where sexually active adolescents aged 15–19 and young adults aged 20–24 are at higher risk for getting STDs rather than adult, and the lag time between infection and complications. Often, a long interval, sometimes years, occurs between acquiring an STD and recognizing a clinically significant health problem [7, 8].

Although most HPV infections resolve on their own and most precancerous lesions resolve spontaneously, there is a risk for all women that HPV infection may become chronic and precancerous lesions progress to invasive cervical cancer [9]. According to the World Health Organization (WHO), cervical cancer will develop after 15–20 years in women with normal immune systems but 5–10 years for women with weakened immune systems, such as those with untreated high-risk HIV infection.

2.2 *The Early Detection and Medical Treatment Interventions*

2.2.1 HPV Early Detection, HPV Vaccinations, and the Medical Treatments

This research applied a Papanicolaou examination among other methods of early detection, for both conventional cytology and liquid based, which is combined with the examination of HPV DNA. According to the NCCN Guidelines ver1.2011 Cervical Cancer Screening, early detection of cervical cancer by Pap smear cytology begins when a woman reaches the age of 21 up to 29 years with the frequency of every 2 years of inspections. For women aged 30 years or more, in addition to cytology, it is also suggested that they undergo examination of HPV DNA. If there is a negative result in cytology and HPV DNA examination, then the examination can be repeated after 3 years [10]. If people are vaccinated with the new HPV vaccines, the rates of infection may be greatly reduced. For now, HPV treatment focuses on the symptoms of the infection. Symptoms include genital warts associated with low-risk HPV types (which don't generally lead to cancers) and the precancerous changes sometimes associated with the high-risk types of HPV.

When the HPV infection has caused abnormal cell changes that could lead to cervical cancer, there are four main treatment options (http://www.webmd.com/sexual-conditions/hpv-genital-warts/hpv-treatment-is-there-hpv-cure), e.g., (1) watch and wait (sometimes the cell changes (cervical dysplasia), precancerous cell changes, or cervical intraepithelial neoplasia will heal on their own), (2) cryotherapy involving freezing the abnormal cells with liquid nitrogen, (3) conization removing the abnormal areas, and (4)] LEEP or loop electrosurgical excision procedure. The abnormal cells are removed with an electrical current. The goal is to remove all the abnormal cells and thus remove most or all of the cells with HPV.

2.2.2 Government Intervention in Cancer

The research on government intervention in cancer by Indra [11] explored the Effectiveness of Educational Intervention of Women's Participation in Cervical Cancer Screening by Acetic Acid Application on the Cervix Versus Pap Smear for Screening Precancerous Cervical Lesions. It aims to evaluate the effectiveness of health intervention on the knowledge of women regarding early detection and prevention of cervical cancer and to determine the association between the selected demographic variable and the knowledge of women on cervical cancer. An experimental approach by knowledge intervention has been applied. Praditsitthikorn, Naiyana et al. [12] analyzed the economic valuation of policy options for the prevention and control of cervical cancer in Thailand to identify the optimum mix of interventions that are cost-effective, from societal and healthcare provider perspectives, for the prevention and control of cervical cancer implemented a model-based cost-utility analysis. Buehler, Sharon K [13] proposed the effectiveness of a call/recall system in improving compliance with cervical cancer screening: a randomized controlled trial. The objective is to determine the effectiveness of a simple call/recall system in improving compliance with cervical cancer screening among women not screened in the previous 3 years. An experimental research, and prospective randomized controlled study, has been applied.

2.2.3 Agent-Based Modeling Approaches for Government Intervention in Healthcare

Several researches on agent based have been developed in the past 10 years to demonstrate the government intervention in healthcare. Michael J. Widener [14] proposed an agent-based modeling of policies to improve urban food access for low-income population by an agent-based modeling—food shopping behavior. This research explores the food purchasing behavior of low-income residents in Buffalo, New York, USA, to simulate the impact of various policy interventions on low-income households' consumption of fresh fruits and vegetables. Bertha Maya Sopha et al. [15] explored policy options for a transition to sustainable heating system diffusion using an agent-based simulation to identify potential interventions for the uptake of wood-pellet heating in Norway using an agent-based model (ABM). Agent-based modeling—heating system. Marina V. Sokolova et al. [16] developed a model and implemented an agent-based environmental health impact decision support system as an approach to the creation of an agent-based system for the assessment of environmental impact upon human health.

Agent-based environmental health DSS. Jiang Wu et al. [17] proposed a multi-agent simulation of group behavior in e-government policy decision simulation. It proposed a multi-agent qualitative simulation approach using EGGBM (E-Government Group Behavior Model) to analyze it from the perspective of system. Agent-based E-Government Group Behavior Model.

From the authors' point of view, there has been no research in cervical cancer which exposed the involvement of multidimensional factors such as interaction among agents, behavior, and the involvement of all agents due to the issue, and no similar tabletop exercise has been implemented as an alternative on decision support system of the reducing policy of cervical cancer at present.

3 Methodology

Observations and preliminary study are carried out by means of direct visits to Dr. Cipto Mangunkusumo Hospital, Jakarta. The aim was to recognize the cancer care (in general), especially through the decision of treatment for the medium to the last stage. Management board interview was carried out to identify what is the most crucial issue on the decision-making of cancer care especially related to the national issue to identify the problems on national cervical cancer policy for the importance and significance statement of this study.

Referring to the observations and preliminary studies, a global picture of the problems was provided. The issue on how to implement a better policy on solving the spread of cancer in Indonesia due to the limited national resources (including national budget) is essential.

The semi-structured interview was conducted related to cervical cancer, which includes the medical practitioners at Radiotherapy Department at Dr. Cipto Mangunkusumo Hospital as the first national reference hospital in Indonesia. Afterward, an interview with "Direktorat Jenderal Bina Pengendalian Penyakit dan Penyehatan Lingkungan/PPPL SubDit Penyakit Kanker" (Directorate General of Disease Control and Environmental Health sub-directorate Cancer) was also carried out.

The secondary data are collected mostly from the Ministry of Health database and other reports related to the government intervention in cervical cancer.

The agent-based simulation building was carried out by implementing Gilbert [18] (1) identification of the target system, (2) design of the simulation, (3) design of the interaction among agents, and (4) the validation. In this research, the target system is the spread of cervical cancer by high-risk human papillomavirus. Refining from the general research topic into specific research objectives and defining the scope of the model are significant. A simple preliminary model was carried out, and then an effort was taken to develop better model representing the real phenomenon of cervical cancer spread. Afterward, the researcher described all type of objects required in the simulation, both agents and its environment [19]. A hierarchical class was implemented to classified sub-objects and then specified the attributes needed as the characteristic of a feature or an object to design interaction among objects. The validation for this simulation considered internal validation and external validation. The internal validation assessed by verifying that its data, variables, and parameters are based on experimentally developed theories. This validation considers the conceptual validity check (is it valid to serve its purpose or not) and verification process. During the external validation process, the test of model's accuracy (of how close the model can reflect the reality) was implemented. An intensive focus group discussion was conducted involving the top management to improve and validate the model.

4 Simulation Building

This model simulates the spread of the human papillomavirus (HPV), which only occurred via sexual transmission. This model also illustrates the effects of certain sexual practices across a population. Adopting NetLogo AIDS model library by Uri Wilensky [20] which has a similar behavior on spread behavior, the model framework was developed.

4.1 Basic Assumptions in Model

For building the agent-based model, in this study several assumptions are applied, as follows:

1. The transmission of HPV is the main factor of cervical cancer; other multifactors (early pregnancy, oral contraceptive use, and smoking) are ignored. The justification is that human papillomavirus (HPV) is the most common cause of cervical cancer [21].
2. The sexual behavior is only assumed to be implemented by sexually active agents over 15 years old and a heterosexual only. The probability of women <15 years applying sexual activity or any sexual criminology is ignored. The sexual relationship between homosexual and agents is also ignored. A couple assumed to always have sexual relationship. The possibility of a couple not applying sexual relationship is also ignored.
3. Women agents will give birth after 2 years of coupling, or there are 5% of the population giving an unplanned birth.
4. Each pregnant woman will only limit to have a maximum of two children. The possibility of women to have only one child or having more than two children is ignored.
5. The differences of age among agents are distributed randomly.
6. If an agent realized that he/she is infected, the agent will not infect the HPV to his/her partner.
7. People live between 0 and 80 years and died afterward. The possibility of people who die between 0 and 80 years is ignored.
8. The HPV transmission occurred to all agent, but the cervical cancer is only for women; even when the men agents are infected by high-risk HPV, there is a possibility of men to suffer other types of cancer, but considering this study which only discussed on cervical cancer, other cancers are ignored.

4.2 Agent Specifications

In this research, agents are defined by gender specified by "person" and "business person" shapes. The assumption is that the "person" shape is identified as woman, and the "person business" shape is identified as man. It also assumed that woman is the one who first decide on mating or not.

4.3 Agent Attributes

The agents' attributes, type, and descriptions are listed in Tables 1 and 2. Those informations show whether the agents are infected or not, knowledge about the infection, duration of the infection, the couple's condition, awareness level, and other demographic condition.

4.4 Environment Specification Where Agents Interact

Agents (men and women) move randomly and engage in a "random walk." Each agent walks one step away from its current location in a different random direction at each clock tick. This movement is known as walking a 360-gon "lattice." The model uses "couples" to represent two people who engaged in sexual relations. Individuals wander around the world when they are not in couples. Upon coming into contact with a suitable partner, there is a chance the two individuals will "couple." When

Table 1 Man attributes

Attribute	Type	Description
Infected?	Booelean	Either a person is infected or not, True for Yes and False fo No
Known?	Booelean	Either a person know about the infection or not
Infection–length	Number	How long the person has been infected
Coupled?	Booelean	Either a person is in a sexual relationship or no
Couple-length	Number	How long the person has been coupling
Commitment	Number	How long the person will stay in a couple relationship with the current partner
Coupling-tendency	Number	How likely the person is to join a couple
Partner	Number	The person that is our currentpartner in couplehood
Age	Number	Age of person in week
Awareness	Number	Percentage of awareness of a person
Unprivileged?	Booelean	Whether a person is poor or not

Table 2 Woman attributes

Attribute	Type	Description
Infected?	Booelean	Either a person is infected or not, True for Yes and False fo No
Known?	Booelean	Either a person know about the infection or not
Infection–length	Number	How long the person has been infected
Coupled?	Booelean	Either a person is in sexual relationship or no
Couple-length	Number	How long the person has been coupling
Commitment	Number	How long the person will stay in a couple relationship with the current partner
Coupling-tendency	Number	How likely the person is to join a couple
Partner	Booelean	The person that is our current partner in couplehood
Age	Number	Age of person in week
Awareness	Number	Percentage of awareness of a person
Unprivileged?	Booelean	Whether a person is poor or not
Vaccine-visit	Number	How many time a person visit a healthcare faciities for vaccination for total immunity
Vaccine-visit-length	Number	How many weeks she does the 1st vaccine immune
Cyt-length	Number	How long a CIN stage in week
Visit-test-time	Number	How long she has been visited the last test
Times-test	Number	How many times a woman got tested

this happens, the two individuals no longer move about; they stand next to each other holding hands as a representation of two people in a sexual relationship with a gray background in the monitor to symbolize coupling.

The presence of the virus in the population is represented by the colors of individuals. Green, yellow, blue, and pink to show whether agents' condition (for men or women) is normal, consciously infected, unknowingly infected, or immune to HPV, respectively. Red, orange, and magenta are representing women with a cervical cancer, consciously CIN+, or unknowingly CIN+, respectively.

The couple also can separate, depending on the rate of each person's commitment attributes. If an infected agent was coupling, then the partner would be certainly infected, unless both of them are immune due to the early vaccination, and when they separated, an agent moves to another agent and infected the new partner (unless the new agent is immune). If an agent was infected, there is also a chance to do the medical treatment, based on her own awareness to the treatment. This model also accommodates agents' cycle of birth and death. Agent with age under 15 years old is assumed to stay with his or her parents.

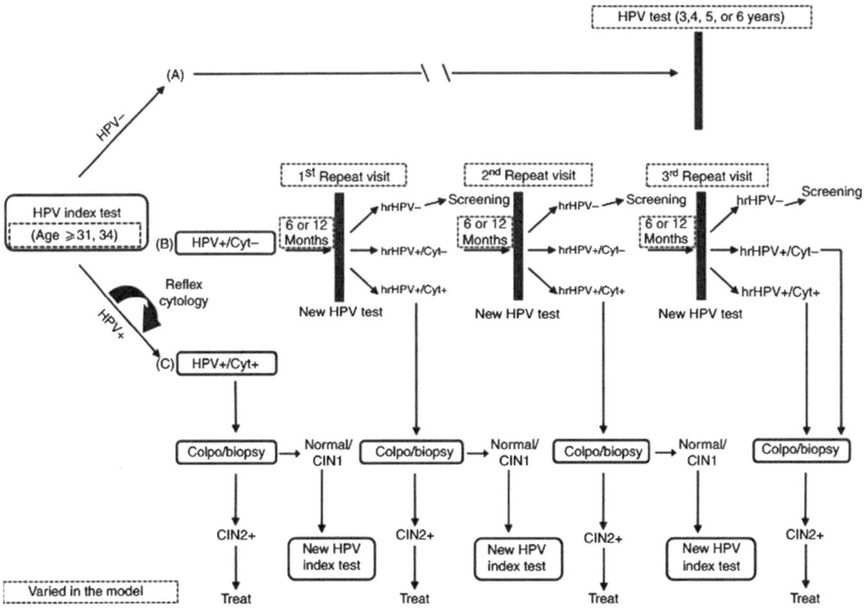

Fig. 1 Flow diagram for proposed HPV DNA screening strategy (Source British Journal of Cancer [22] 106, p. 1574)

4.5 Decision Rule Specification and Interaction Among Agents

The following figure shows the decision rule implemented in this model due to the HPV test/screening and treatment for cancer. These decision rules are adopted from Norwegian government in which the HPV test could be implemented due to the availability of budget (Fig. 1).

4.6 HPV Spread Model

An agent can be in one of the three states: either healthy or not infected by HPV, consciously infected, or unknowingly infected. The medical resources stated that mostly there is 90% of HPV infection chances during the first coupling. The model represents sexual activity as two people standing next to each other. This indicates that all couples have sexual activity and practice a singlehood. There is also a chance that people are coupling without having sexual relationship, but due to simplicity it was ignored. The couple can be separated, depending on the rate of each person's commitment attributes. If an infected agent was coupling, then the partner would be

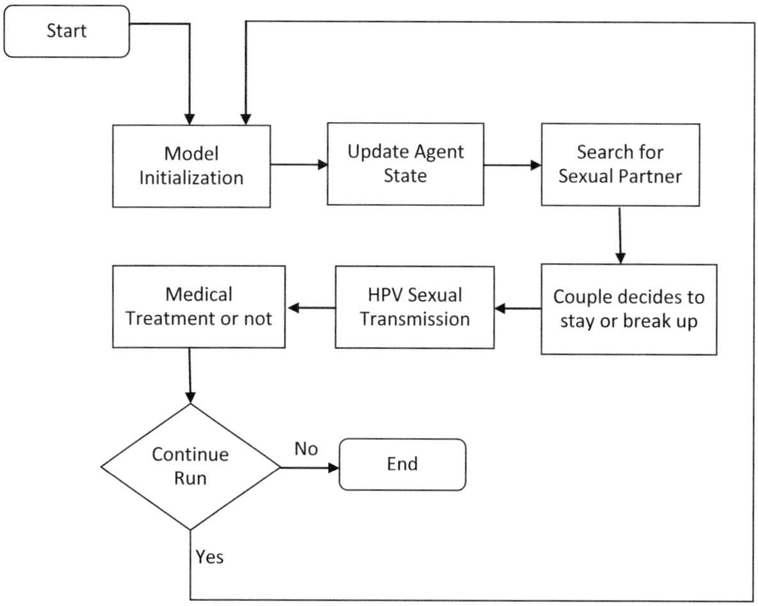

Fig. 2 Flowchart of HPV spread model

certainly infected, unless both of them are immune due to the early vaccination, and when they separated, an agent moves to another agent and infected the new partner (unless the new agent is immune).

If an agent was infected, there is also a chance to do the medical treatment, based on his/her eagerness to the treatment, and due to simplicity 100% treatment leads to healthy condition/not infected. This model also considers the evolution situation (breed) (Fig. 2).

4.7 Validation

The internal validation is carried out by a group discussion with oncologist experts based on compliance to the theory of HPV and cervical cancer. Several parameters of correction have been completed due to the decision rules and new variable of the possibility that a person has the capability to clear the HPV within 8 months after the infection. Due to the lack of research in the probability of Indonesian immunity to this virus, the model provides a slider to its ability in changing the value to consider as an external condition during the experimental process. Thus after the last confirmation with the experts due to this matter, finally validity check was confirmed to answer its purpose. As the test of model's accuracy of how it reflects the reality on the cervical cancer spread, external validation has been carried out. Based on Schmid [23], this model checked for:

1. *Correspondence theory of truth*: The agents in the conceptual model consist of two agents, namely, woman and man, whose attributes are also in their own real life.
2. *Consensus theory of truth*: The validity was checked for the extreme condition where if the commitment is 0, the coupling tendency 10, and the awareness 0, the simulation shows the increase of HPV infection among women, and in a thick of 120 years, the population vanished since every woman suffers for cancer; thus, no woman can give birth.
3. *Coherence theory of truth*: The simulation is adopted from cervical cancer theory, confirmed by medical experts. Thus the theory of HPV transmission to cancer is believed to be true.

The agent-based model accuracy can be classified into four levels (Epstein and Axtell [24]). In this level of accuracy, the evolution in the model is similar with the evolution in the real world. The model accuracy of this research is significant with the Level 1 of a qualitative agreement with the empirical macrostructure. To be significantly in quantitative agreement with empirical macro- and microstructure further research, the Level 2 and Level 3 of model accuracy are not able to be done before the recommended future research is implemented, considering no research due to the empirical macrostructure has been conducted in accordance to the limited budget and human resource.

5 Simulation Result

The free HPV test/screening test slider and the free vaccination slider indicate government interventions, which may provide a reasonable solution to this problem. The vaccination of HPV as stated in the basic assumptions is only provided for those aged 9 to 15 in this model. The assumption is that the vaccinated young woman will have immunity to HPV. These two sliders illustrate as one of the early detection and prevention programs done by the government. The HPV test in this model is adopting the rule of HPV testing in Norway [10].

By implementing vaccination to the right target, thus the impact on decreasing the infection spread can be achieved. The eagerness of a person to get a HPV medical treatment when the result is HPV+ and CIN+ will also give the individual the cure to stay healthy which will affect the decreasing of infection spread.

The free treatment sliders (per stage) indicate the government intervention in treatment program in whether possible or not the unprivileged will be freely benefited from the free cost of cervical cancer treatment. The issue occurred is whether the government is able to deal with the limited budget or not.

This simulation has been constructed to answer the first research question of the mechanism to describe the sexual network to have significant implication on the

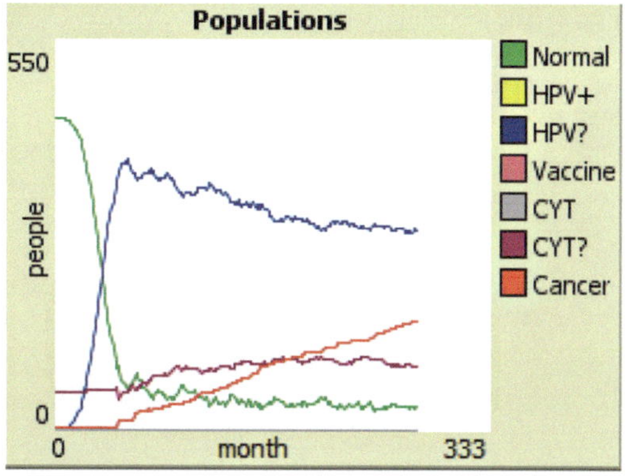

Fig. 3 The modeling result 1 if no intervention occurred

spread of high-risk HPV epidemic. The spread of HPV declined with the increase of commitment or the tendency of free sexual behavior. The spread of HPV also declined with the decline of couple tendency. If most people prefer to stay single rather than coupling, a minimum transmission will occur.

To answer the second question of how will it happen if no interventions occurred, this simulation is consistent with the theory that the sexual behavior is increasing in adolescent (http://www.o-bras.com/2010/05/data-survei-seks-pelajar-indonesia_3019.html). For an average coupling tendency of 5 (from a scale of 10), the result on Fig. 3 demonstrated that the high-risk HPV infection will spread widely and rapidly and will decrease the healthy people in 5–10 years ahead (shown by the green line). There is also a tendency of an increase in the population of unknowingly infected HPV which also increases the amount of cervical cancer.

What will happen if interventions occurred (the sexual health promotion, the HPV vaccinations (national program), and the medical treatment for patients with HPV+)? Due to promotion of healthy sexual life intervention, Fig. 4 showed significant findings. The intervention will not hold the infection spread if it was carried out partially. Figure 4 showed an increasing amount of both people with unknowingly high-risk HPV and spread of cervical cancer and also a decreased pattern of healthy people if only a partial intervention is implemented. The intervention of high-risk HPV awareness only will restrain the amount of unidentified patient with high-risk HPV (the blue line).

But if the interventions are carried out simultaneously, a better pattern can be generated. Figure 5 showed how interventions on sexual behavior, high-risk HPV awareness, and free early detection will give a better outcome rather than implementing only intervention on sexual behavior and high-risk HPV awareness.

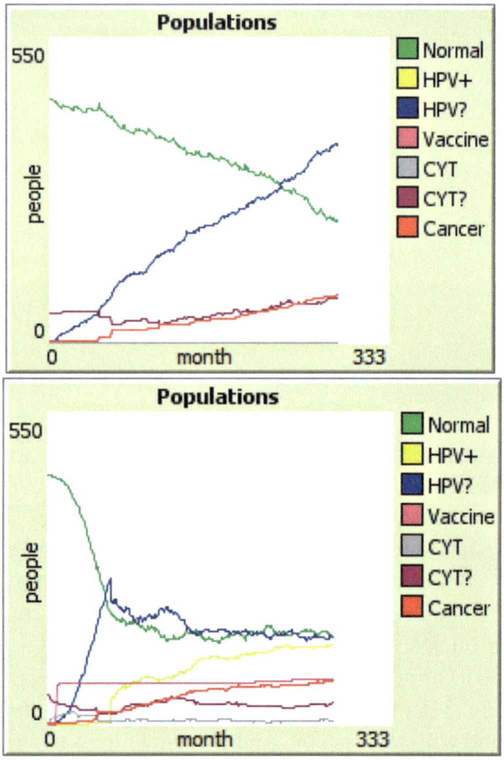

Fig. 4 The modeling result 2 and 3 intervention on sexual behavior only and on high-risk HPV awareness only

6 Summary and Conclusion

6.1 Conclusion

This simulation describes the phenomena of HPV spreading in a local area on how the spreading is very dependent on the commitment of the population in their sexual behavior, the distribution of age to implement vaccine program, and the awareness of people to go to the healthcare facility. When an economical status takes into account, people are not able to go to the healthcare facility due to their limited financial budget. The result of this research represented the impact of one intervention to another, in terms of how many people suffer from cancer, how many people die by cancer, and how many people is infected by the HPV (whether they have realized that they were infected or not).

The experimental occurred to observe the possible intervention, and a recommendation has been proposed to provide the government with the alternative solution on whether the policy to free of charge in HPV test/screening, HPV vaccine, or cervical cancer treatment is possible to implement or not. Regarding the expanded budget on these interventions, this research demonstrated a significant intervention

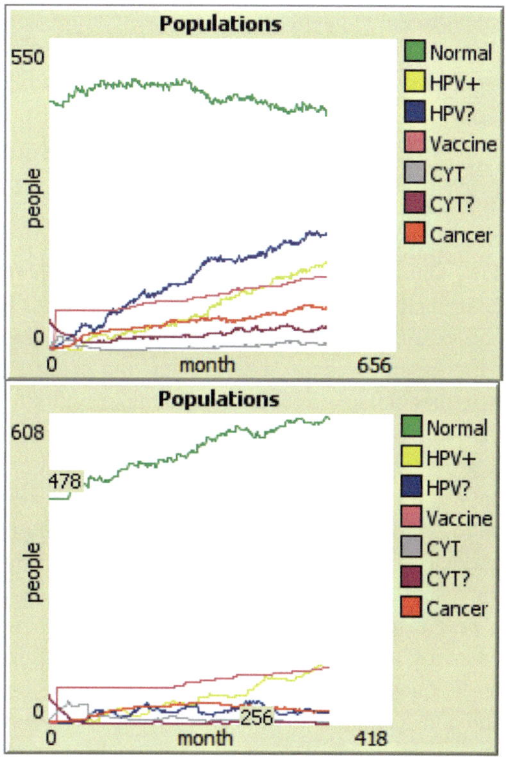

Fig. 5 The modeling result 4 intervention on both sexual behavior and high-risk HPV awareness and the modeling result 5 intervention on sexual behavior, high-risk HPV awareness, and free early detection

on specific promotions which would increase the coupling commitment. This study presented an agent-based simulation approach for illustrating the HPV infection spread by sexual behavior. Significant findings occurred in the sensitivity analysis to illustrate the interventions. Policy simulations are valuable because of their ability to facilitate training and feedback concerning potential impact of decisions. Policy simulation models are useful because they accelerate creation of scenarios, allow rapid changes in parameters, and provide a test bed for concepts and strategies. The limitation of the types of interventions also needs to be underlined because our further research should consider interventions which will be more in line with the socioeconomic conditions of the Indonesian community and the government budget for the provision of vaccines, for instance.

6.2 Implications

The implication of this research enables policy makers to model complex, adaptive social systems and their associated policy problems even when data are limited on information diffusion through social networks of actors, public opinion formation, institutional behavior, and decision-making of how a HPV is spread in a variety of ways such as sexual contact. The research presented here is an initial attempt to model these processes and their effects on public health.

6.3 Future Directions of Research

This simulation proposed as a tool to help the decision-makers understand the complexity of the problem broadly; it captures the complex network of interactions and connections that make up real systems and make it possible to see emergent patterns and unexpected changes and events. When it leads to government, the experimental is possible to implement without any risk of real implementation in the society. This model gives insights into causes of emergent phenomena and it provides framework. Agent-based simulation provides a flexible framework for answering research questions, i.e., why is this happening, what if these trends continue, what will happen next, what is the best scenario that can happen. This cervical cancer simulation will open a new opportunity of new research regarding the research and measurement of the model to compare with the reality. It also opens a new research on measuring "Indonesian average commitment on their sexual behavior" or modeling promotion strategy to increase public average commitment on Indonesian sexual behavior and also Indonesian awareness as the very important and complex parts to define.

This research also opens collaboration among parties, between academician in management and experts in oncology sectors to have the same spirit on reducing the mortality rate or incidence rate.

References

1. Pisani, P., Parkin, D. M., Bray, F., & Ferlay, J. (1999). Estimates of the worldwide from the 25 cancers in 1990. *International Journal of Cancer, 83*, 18–29.
2. Schellekens, M. C., Dijkman, A., Aziz, M. F., Siregar, B., Cornain, S., Kolkman-Uljee, S., Peters, L. A. W., & Fleurena, G. J. (2004). Prevalence of single and multiple HPV types in cervical carcinomas in Jakarta, Indonesia. *Journal of Gynecologic Oncology, 93*, 49–53.
3. Sibuea, W. H. (2000). Cancer in Cipto Mangunkusumo Hospital Jakarta. *Gan to Kagaku Ryoho, 27*, 549–559.
4. Tjindarbumi, D., & Mangunkusumo, R. (2002). Cancer in Indonesia, present and future. *Japanese Journal of Clinical Oncology, 32*, 17–21.
5. Institute of Medicine. (1997). In T. R. Eng & W. T. Butler (Eds.), *The hidden epidemic: Confronting sexually transmitted diseases*. Washington, DC: National Academies Press.
6. Alam, S. J., Meyer, R., & Norling, E. (2000). *A model for HIV spread in a South African village*.
7. Chandra, A., & Stephen, E. H. (1998, January–February). Impaired fecundity in the United States: 1982–1995. *Family Planning Perspectives, 30*(1), 34–42.
8. Colizza, V., Barthélemy, M., Barrat, A., & Vespignani, A. (2007). Epidemic modeling in complex realities. *Comptes Rendus Biologies, 330*, 364–374.

9. Burd, E. M. (2003). Humann papillomavirus and cervical cancer. *Clinical Microbiology Reviews, 16*(1), 1–17.
10. National Comprehensive Cancer Network (NCCN). (2011). *Guidelines clinical practice guidelines in oncology*. Cervical cancer screening. Version 1. http://www.nccn.org
11. Indra, V. (2013). Effectiveness of educational intervention of women's participation in cervical cancer screening by acetic acid application on the cervix versus Pap smear for screening precancerous cervical lesions. *International Journal of Nursing Education, 5*(1), 231–235.
12. Praditsitthikorn, N., et al. (2011). Economic evaluation of policy options for prevention and control of cervical cancer in Thailand. *PharmacoEconomics, 29*(9), 781–806.
13. Buehler, S. K. (1997). Effectiveness of a call/recall system in improving compliance with cervical cancer screening: A randomized controlled trial. *Canadian Medical Association Journal, 157*(5), 521–6.
14. Widener, M. J. (2013). Agent-based modeling of policies to improve urban food access for low-income populations. *Applied Geography, 40*, 1–10.
15. Sopha, B. M., et al. (2011). Exploring policy options for a transition to sustainable heating system diffusion using an agent-based simulation. *Energy Policy, 39*(5), 2722–2729.
16. Marina V. Sokolova, et al. (2009). Modeling and implementing an agent-based environmental health impact decision support system. *Expert Systems with Applications, 36* (2, Part 2), 2603–2614.
17. Wu, J., et al. (2008). Multi-agent simulation of group behavior in E-Government policy decision. *Simulation Modelling Practice and Theory, 16*(10), 1571–1587.
18. Gilbert, N. (2008). *Agent-based model*. Los Angeles: SAGE Publication.
19. Axtell, R. (2000). *Why agents? On the varied motivations for agent computing in the social sciences*.
20. Wilensky, U. (1999). Netlogo. http://ccl.northwestern.edu/netlogo/. Center for connected learning and computer-based modeling. Evanston: Northwestern University.
21. Faridi, R., Zahra, A., Khan, K., et al. (2011). Oncogenic potential of human papillomavirus (HPV) and its relation with cervical cancer. *Virology Journal, 8*, 269.
22. Burger, E. A., Ortendahl, J. D., Sy, S., Kristiansen, I. S., & Kim, J. J. (2012). Cost-effectiveness of cervical cancer screening with primary human papillomavirus testing in Norway. *British Journal of Cancer, 106*, 1571–1578.
23. Schmid, A. (2005). What is the truth of simulation? *The Journal of Artificial, Societies, and, Social,Simulation, 8*(4), 7.
24. Epstein, J. M., & Axtell, R. L. (1996). *Growing artificial societies: Social science from the bottom up*. Cambridge, MA: MIT Press.
25. HPV treatment and cure. http://www.webmd.com/sexual-conditions/hpv-genital-warts/hpv-treatment-is-there-hpv-cure
26. Human Papilloma Virus (HPV) and Cervical Cancer. http://www.who.int/mediacentre/factsheets/fs380/en/
27. http://hpv.emedtv.com/hpv/hpv-and-cervical-cancer-p2.html
28. http://healthypeople.gov/2020/topicsobjectives2020/overview.aspx?topicid=37
29. http://www.o-bras.com/2010/05/data-survei-seks-pelajar-indonesia_3019.html

Method for Getting Parameters of Agent-Based Modeling Using Bayesian Network: A Case of Medical Insurance Market

Osamu Matsumoto, Masashi Miyazaki, Yoko Ishino, and Shingo Takahashi

Abstract To date, agent-based social simulation (ABSS) is a popular method to study the behavior of a social system and the interaction of the constituent members of the system. With the development of computer and information technologies, many ABSS approaches have been proposed with wide application. However, the definitive methodology for modeling of the agent's behavior in ABSS has not been established yet. This study proposes a new methodology of modeling of the agent's behavior in ABSS using Bayesian network based on the questionnaire survey. This method enables us to simultaneously perform the construction of the agent's behavior model and the estimation of the internal parameters within the model. This study took a Japanese medical insurance market as an example, since this complicated market deserves detailed consideration. We verified the effectiveness of the proposed methodology by applying the scenario analysis to this case.

Keywords Agent-based social simulation • Behavior model • Bayesian network

1 Introduction

Computer simulation of social phenomenon is a promising field of research at the intersection of social and mathematical science [1]. Agent-based social simulation (ABSS) is a major computational modeling approach for social simulation. Many social processes were studied using ABSS, including disease, trade, wealth, war, and communication [2]. In real human societies, there are many nonlinear interactions between the members. Although traditional mathematical models can hardly represent such interactions in complex systems, ABSS is a good tool to treat them. ABSS contributes to two aspects: (1) the rigorous testing, refinement, and extension

O. Matsumoto (✉) • M. Miyazaki • S. Takahashi
Graduate School of Creative Science and Engineering, Waseda University, Tokyo, Japan
e-mail: osamu-036@moegi.waseda.jp; masashi-728@asagi.waseda.jp; shingo@waseda.jp

Y. Ishino
Graduate School of Innovation & Technology Management, Yamaguchi University, Ube, Japan
e-mail: ishino.y@yamaguchi-u.ac.jp

U.S. Putro et al. (eds.), *Agent-Based Approaches in Economics and Social Complex Systems IX*, Agent-Based Social Systems 15, DOI 10.1007/978-981-10-3662-0_4

of existing theories that have proved to be difficult to formulate and evaluate using standard mathematical tools and (2) a deeper understanding of fundamental causal mechanisms in a complex system [3]. However, the definitive methodology for modeling of the agent's behavior in ABSS has not been established yet. Researchers usually model each case empirically using a trial-and-error approach. This is mainly because two difficulties exist in modeling of the agent-based model: difficulty in constructing the behavior model of agents and difficulty in estimating the internal parameters within the behavior model. These two problems are mutually related.

Against these problems, we proposed a new methodology of modeling of the agent's behavior in ABSS using Bayesian network based on the questionnaire survey. The main point of the proposed method is that it enables us to simultaneously perform the construction of the agent's behavior model and the estimation of the internal parameters within the model. We set the medical insurance market as a case study. Regarding this target market, we built the Bayesian network model that reflected the realities of consumer's behavior. Finally, we executed the agent-based simulation and investigated the results derived from each scenario.

2 Related Works

2.1 Virtual Grounding Method

The virtual grounding method is a method that was proposed to construct valid facsimile models in ABSS under conditions where the real-world data is necessary to build the behavioral model, but it cannot be obtained [4]. When modeling a social system, people make a model to abstract the target real world, and simultaneously they use data that is sampled from the real world in order to estimate parameters of the model. In ABSS the constructed model produces the simulated data, which is used to explain the phenomena in the real world. This is the process of the general grounding in ABSS. By contrast, the virtual grounding method does not directly use the real-world data. The virtual grounding method needs a questionnaire survey in which participants are asked their behavior under a certain situation. That is, the participants in the survey are used as agent samples, and their answers are used as virtual behavior to estimate parameters of a model that is already constructed before performing the survey. When questions in the questionnaire survey are designed to simulate the changes of a situation, some parameters can be obtained as dynamic parameters, which are recalculated at each access time, i.e., these parameters act as a function. This is one of the main advantages of the virtual grounding method. However, the virtual grounding method can be used only when a certain model has already constructed and accepted as a reasonable model. This method cannot be used when you make a model from scratch, like a case in this paper.

2.2 *Bayesian Network and Multi-agent Simulation*

A Bayesian network is a machine learning technique for empirically identifying associations in complex and high-dimensional data. A Bayesian network graphically models probabilistic relationships among a set of variables. Over the last decade, Bayesian networks have become a popular representation for encoding uncertain expert knowledge in expert systems [5]. More recently, researchers have developed methods for deriving Bayesian networks from data [6, 7].

A Bayesian network can represent nonlinear covariation between features and does not assume a Gaussian distribution of variables, since the relation between features is given in the conditional probability tables (CPT). The model's structure can be learned from the data based on information criteria such as Akaike's information criterion (AIC) and minimum description length (MDL) [8]. The Bayesian network structure obtained from learning enables us to compute the posterior distribution of variables when other variables (the evidence variables) are observed. This process is called probabilistic inference, in which some efficient computational algorithms have been proposed, such as loopy belief propagation [9, 10].

To date, there are a few research works that combine the Bayesian network with the multi-agent simulation: a behavior model by which the utility of a new address is calculated when moving [11] and a behavior model which determines what means of transportation should be taken when a natural disaster occurs [12]. These researches proposed the behavior model that calculates the utility of an agent's behavior using a Bayesian network where the influence of the extrinsic factors is stochastically estimated. The positive side of these researches is that they defined a model which evaluates alternatives of an agent's behavior. However, the existing researches focused on appropriateness of the evaluation of an agent's decision-making, and so parameterization method of an agent has not been discussed in detail.

Our previous research combined the Bayesian network with ABSS, where the Bayesian network was used to represent consumer's behavior in purchasing the health insurance product and then ABSS was performed to examine the effect of the word-of-mouth communication [13]. Since we constructed our ABSS after we had clarified factors that have a strong impact on the purchase behavior, the validity of the behavior model was ensured. At that time, however, the behavior model using Bayesian network did not directly connect with ABSS. In consequence, the parameterization was not enough to be able to simulate various situations.

3 Proposed Method for Getting Parameters of ABSS Model

We proposed a new methodology to obtain parameters of the behavior model of an agent in ABSS using Bayesian network. This methodology enables us to simultaneously perform the construction of the agent's behavior model and the estimation of the internal parameters within the model.

The procedures of the proposed method are as follows:

Step 0 Preparation
We select the target social system and roughly design the agents that constitute the system.

Step 1 Hypothesis Creation
About the selected system, we shed light on all of the factors that relate to an agent's behavior, using existing data and/or the experiences of experts. Then, we create the hypothesis of the agent's behavior model, in which the behavior is interpreted as cause-and-effect relationships of factors. The hypothesis about the agent's behavior model can be expressed as a network structure of the factors.

Step 2 Agent-Based Model Design
We design the architecture of the ABSS targeted. Based on the created hypothesis about the agent's behavior model, we design an agent's behavior in detail using the concrete parameter variables and also design a system world (society). Determining variables that relate to the agent's behavior is very important, although the distribution of the variable values and the precise interactions between variables are unknown in this stage. In this stage we extract the question items that should be asked to the respondents of the questionnaire.

Step 3 Questionnaire Survey
We design and perform the questionnaire survey in order to understand people's behavior regarding the extracted factors.

Step 4 Validation and Update of Behavior Model
We examine the validity of the hypothesis from the obtained survey data. We evaluate the appropriateness of the network structure of the behavior model, which is made in Step 1, using the information criteria and/or chi-square value. If there are insufficient parts in the network structure of the behavior model, we modify them.

Step 5 Parameter Determination
The Bayesian network method calculates the CPT of each factor by applying the survey data to the agent's behavior model in which the structure is finally determined in Step 4. Then, the probabilistic inference is executed to acquire the value of parameters.

4 A Case of Japanese Medical Insurance Market

The insurance industry in Japan has undergone drastic changes, since the new Insurance Business Law became effective in 1996, aiming to loosen regulations on insurance companies. To date, consumers' attitude toward insurance products has greatly changed compared to before. Economic growth has recently slowed as a consequence of the current world economic crisis. And the labor force in Japan is shrinking because of the falling birthrate and the aging population. Accordingly, Japanese consumers came to put a higher value on the existence security that enhances the medical treatment, pension, and nursing, rather than the expensive life insurance against death. Understanding customers' needs and values in purchasing insurance products has become more important for insurance institutions than ever before, because of not only environmental changes but also changes in customer behavior.

In this chaotic market situation, modeling the consumer's product choice behavior is very meaningful to make the marketing strategy about the insurance products. The mass media advertising and the activities of sales representatives are especially important to sell the medical insurance products [14, 15]. An ABSS model should be constructed to make it possible to analyze the situation from these two perspectives.

5 Application to Medical Insurance Market

5.1 Hypothesis Creation About an Agent

As a result of analyzing consumers' decision-making of the purchase of the medical insurance products, it was found that there are two important factors: one is the current purchase status about the medical insurance, and the other is the mood of having people think "it is time for me to purchase the medical health insurance" [16]. We call this mood "timely state of mind." Moreover, the previous research indicated five motivations which solicit for the purchase of the medical insurance: changes in life stages, word-of-mouth communication, anxiety about health, the advertisement, and the insurance renewal [17]. Based on the foregoing knowledge, we created the hypothesis about the agent's purchase behavior of the medical insurance, as shown in Fig. 1. An agent decides to make the purchase or the cancelation of the medical insurance going through some events including changes in life stages, word-of-mouth communication, the advertisement, and so on.

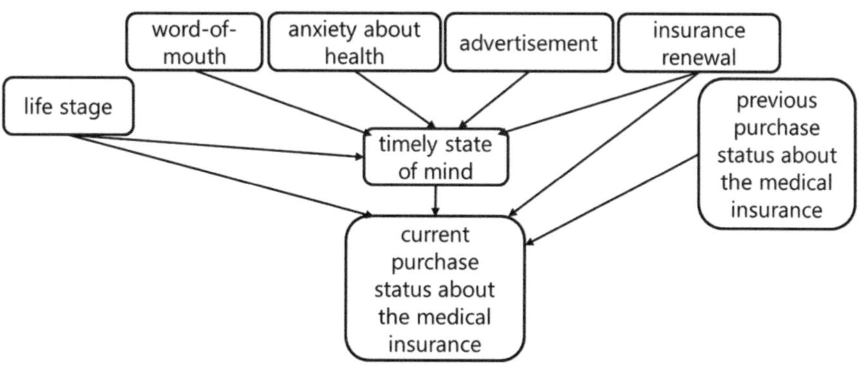

Fig. 1 Hypothetical agent's behavior model of the medical insurance

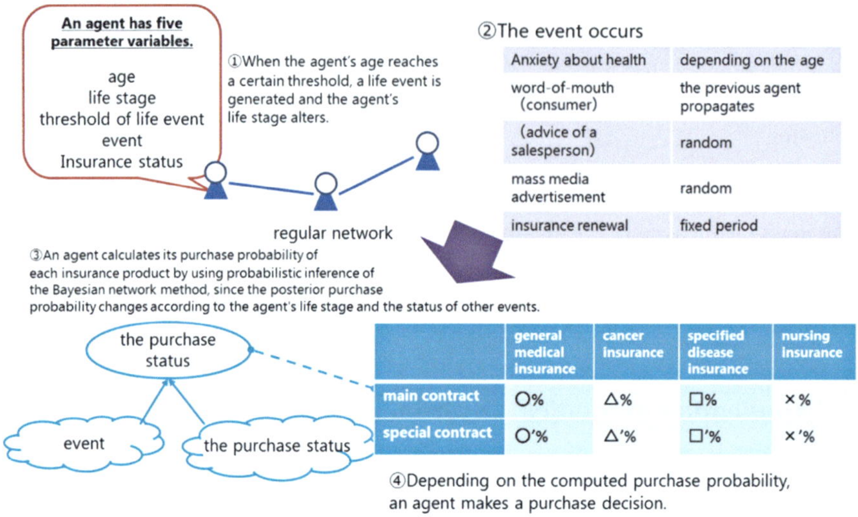

Fig. 2 Conceptual scheme of the model

5.2 *Agent-Based Model Design*

Subsequently, we constructed a regular network of agents as a place or a society where agents interact with each other. Regular networks are "regular" because each node has exactly the same number of links. The feature of a regular network is defined by the number of constituent nodes and the average degree of the network [18]. Edges in a network indicate the close relationships between agents. For example, an event to tell a friend about what an agent experienced can be realized using the edges. Figure 2 shows the conceptual scheme of the agent-based model that we constructed.

An agent has five parameter variables, which were named *age*, *life stage*, *threshold of life event*, *event*, and *insurance status*, respectively, as shown in Fig. 2. The *age* variable represents the agent's age, which increases according to the simulation step. When the agent's age reaches a certain threshold, a life event is generated and the agent's life stage alters. Seven kinds of life events were prepared: heading into the workforce, getting married, giving a birth to a baby, sending a child to college, buying a home, bringing up a child to be independent, and retiring from work. Since it is assumed that these seven kinds of life events occur in this order, eight life stages exist. The *threshold of life event* variable indicates the threshold value of the agent's age, and the life event occurs when the agent's age exceeds the threshold. Each agent has the unique value of the *threshold of life event* variable before starting the simulation. The *event* variable indicates a single-shot event other than the life events described above, which leads to the purchase of the medical insurance product. The *event* variable concretely consists of an array of the state of four kinds of incidents, which are word-of-mouth communication, anxiety about health, the contact to the mass media advertisement, and the insurance renewal. Finally, the *insurance* variable indicates the purchase status on eight kinds of the medical insurance, which are as follows. Since you are given a choice from four types of the medical insurance (general medical insurance, cancer insurance, specified disease insurance, and nursing insurance) and two modes of the contract agreement (the main contract and the special accessory contract), there are eight kinds of the medical insurance, i.e., four times two equals eight.

At each simulation step, an agent calculates its purchase probability of each insurance product by using probabilistic inference of the Bayesian network method, since the posterior purchase probability changes according to the agent's life stage and the status of other events. Then, depending on the computed purchase probability, an agent makes a purchase decision.

5.3 *Questionnaire Survey*

Based on the agent's behavior model derived from the hypothesis, we designed the questionnaire survey. The number of the respondents was determined by the following equation:

$$n = \frac{N}{\left(\frac{C}{2k}\right)^2 \times \frac{N-1}{p(1-p)} + 1}$$

where N is the parent population. As the number of the parent population, we used 103,510,000, the number of 20-year-old or more men and women of Jan. 1, 2014. And C, k, p, and n represent the confidence interval coefficient, the reliability coefficient, the population proportion, and the needed sample number, respectively:

$$n = 666, \text{when } C = 0.1, k = 2.58, p = 0.5$$

In case some respondents might not answer, the number of the samples was decided as 800. The allocation of respondents in terms of sex and age was performed mirroring the Japanese insurance market. The questionnaire survey was performed through the Internet from Aug. 20 to Aug. 25, 2014.

5.4 *Validation and Update of Behavior Model*

A node existing in the graphical model of the agent's behavior signifies one item of the questionnaire. For instance, the anxiety about health, the parameter variable, is expressed by a certain node, which has two states: the state value of the node is 1 when the agent is worried about its health, and the state value is 0 when the agent is not worried about it. As a result of the survey design, some factors possess only one node, but other factors possess more than one node. We call the factor in the latter case a group entity. For example, the *life stage* is a group entity consisting of eight nodes, including “heading into the workforce,” “getting married,” “giving a birth to a baby,” and so on.

We examined the validity of the hypothesis of the agent’s behavior, i.e., the reasonability of the model structure, by obeying the following procedures:

1. The full search for appropriate connections of nodes within a group is performed based on the AIC value.
2. Linkage strength of all the combinations of nodes belonging to different groups, which are connected in the hypothetical behavior model, is investigated by Pearson’s chi-square testing. Then, when arbitrary two nodes significantly correlate to each other, these nodes are connected by an edge (10% significance level was used in this study). The direction of the edge, the direction of cause and effect, is already determined by the hypothesis.
3. The significance of the relations between nodes that do not compose a group is detected by the log-linear model. This method considers interactions between more than two nodes, in addition to the relationships between two nodes. Then, when it is found that the arbitrary two nodes significantly correlate to each other and the connection can be reasonably explained, these nodes are connected by an edge.

These procedures found insufficient parts in the network structure of the hypothetical behavior model, so that they were revised. The consequent behavior model was a little different from the hypothetical behavior model. Figure 3 shows the agent’s behavior model finally modified. As shown in Fig. 3, the edge from “the anxiety about health” to “the media advertising” was newly added. We had not expected the existence of this cause-and-effect relationship. Moreover, it was found that “the anxiety about health” and “the word-of-mouth communication” are directly connected to the purchase status, not going through “timely state of mind.”

5.5 Parameter Determination

The parameter values we should obtain are the probability to possess each insurance product when an agent is placed in a varied situation. In the behavior model shown in Fig. 3, the number of nodes we can operate is 18, because the *life stage* and the *current insurance* finally have 6 and 8 nodes, respectively, and other 4 variables have only one node each. Since all these variables are the binary variable, the number of an agent’s state is 2^{18}. Parameter values were computed in all kinds of an agent’s state, by applying the probabilistic inference of Bayesian network based on the obtained graph structure of the agent’s behavior model. In short,

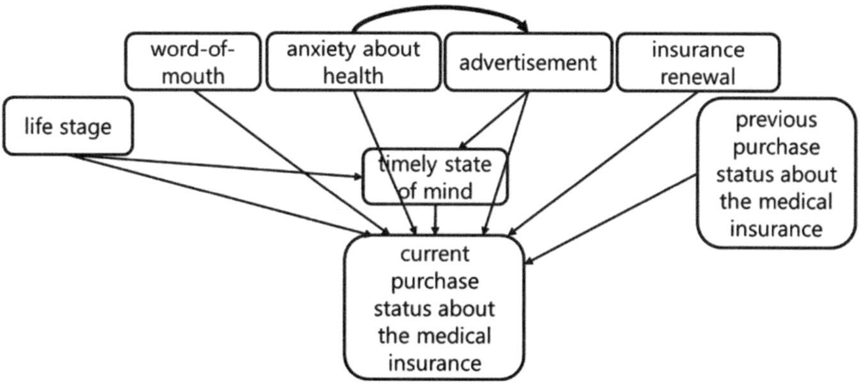

Fig. 3 Agent's behavior model modified

the objective variables, which are the values of the purchase probability of eight kinds of insurance products, were calculated under assumption that the state of 18 explanatory variables can be observed.

6 Computer Simulation Experiments

The simulation experiments were performed by using the agent-based model constructed targeting Japanese medical insurance market.

6.1 Validation of Obtained Parameter Values

In nature, the proposed methodology should be validated by investigating whether the obtained parameter values are reasonable or not. However, since the obtained parameters are a large amount of probability, which are totally different from the results of previous methods, it is difficult to directly examine the reasonability of the obtained parameter values. For the sake of comparison, therefore, we constructed other agent's behavior models derived from the questionnaire survey data simply based on the AIC, with no hypotheses. Then, the simulation experiments were, respectively, executed using the two different behavior models: one model made by the proposed method and the other model made without any hypothesis.

The number of agents was set to 100. A time step in the simulation experiment corresponds to 6 months in the real world. A trial of the simulation requires 10 time steps, which correspond to 5 years in the real world. In each case, 100 trials were performed. Figure 4 shows the ratio of agents that possessed the medical insurance,

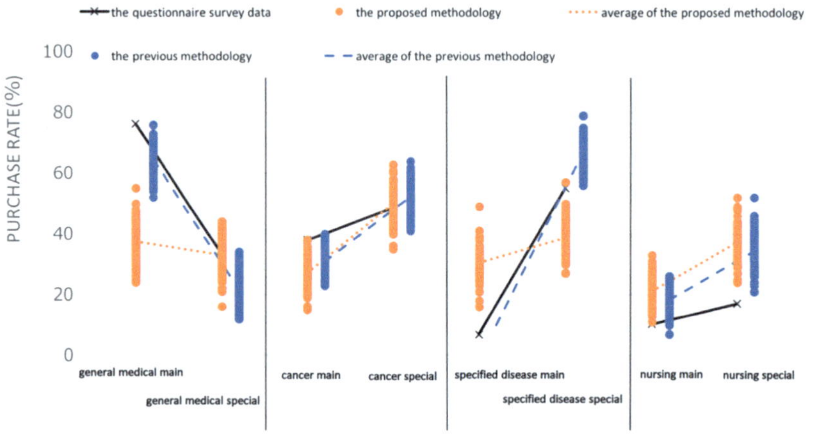

	general medical main	general medical special	cancer main	cancer special	specified disease main	specified disease special	nursing main	nursing special
t-value	8.22515E-85	2.67408E-36	1.08978E-09	0.001695943	2.12924E-89	6.01664E-94	1.07771E-06	0.00013226
t-value(%)	0.00%	0.00%	0.00%	0.17%	0.00%	0.00%	0.00%	0.01%

Fig. 4 Comparisons of the simulation results by the two models

when the simulation finished. Consequently, both models were able to reproduce the similar ratio of agents that possessed each medical insurance product, compared to the questionnaire data. It was found that there was no major difference between the two models.

6.2 *Scenario Setting*

The scenario analysis has been widely used as a means to produce information that will guide decision-making under a variety of alternative futures. We are able to find the possibility of changing the target system by performing the simulation under some scenarios that are composed of a combination of the several alternatives that would trigger the distinctive market configuration and several strategic options that would affect the market situation. We call them a situation alternative and a strategic option, respectively. The abovementioned six kinds of *life events* were set as situation alternatives. The ratio of running an advertisement and the visiting probability of the sales force were set as strategic options. Concretely, two types of selling strategy were prepared: the first strategy invested the budget only in the face-to-face selling, and the second strategy invested the budget in both the face-to-face selling and the media advertisement. There were 12 (= 6 × 2) scenarios in total, since six situation alternatives and two selling strategies were prepared.

6.3 Results of Scenario Analysis

For each scenario, 100 trials of the simulation were performed. The mean values of the penetration ratio of each medical insurance product under each scenario are depicted in Fig. 5. From Fig. 5, the following knowledge about the marketing strategy was found.

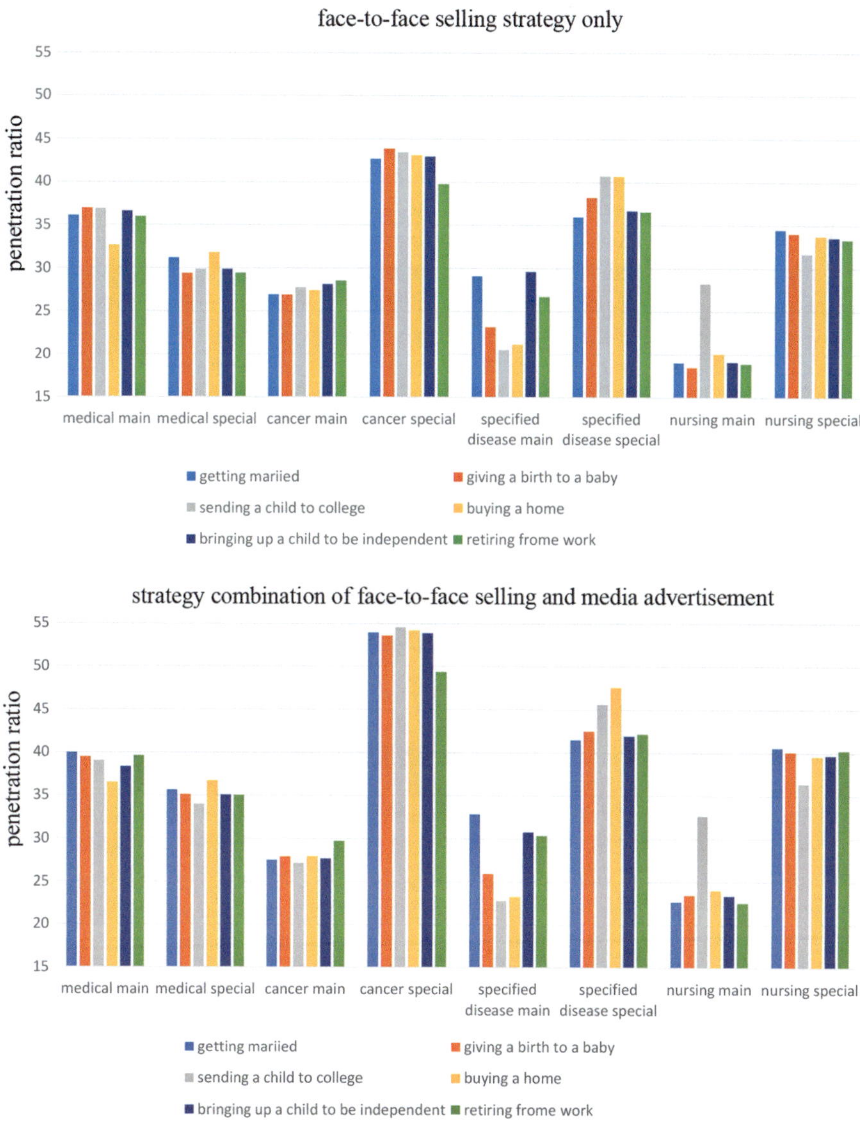

Fig. 5 Penetration ratios of medical insurance products under scenarios

The penetration ratio falls when all marketing budget was invested in the sales activity (the face-to-face selling), without the media advertisement. This occurred as an inevitable consequence, because the resource that an insurance company had spent for the media advertisement in the second scenario was simply adjusted to zero in the first scenario. However, what should be paid attention to is that the main contract type of the cancer insurance alone has little difference of the penetration ratio between the different strategies. In short, in the case of the main contract type of the cancer insurance product, the penetration ratio is little affected even if the media advertisement is stopped. This might be inferred as follows. The cancer had become common but serious disease, and the awareness of the cancer insurance product increased, so that people have come to purchase the cancer insurance product regardless of the presence of the media advertisement. This is a new finding derived from the scenario analysis.

Next, we investigated the penetration ratio of each insurance product according to the *life stage* of consumers. It was found that the penetration ratio of the main contract type of the general medical insurance product is lower under the scenario of "buying a home" than that of other scenarios. Since "buying a home" is often the most expensive shopping in life, people might want to reduce the expense. Consequently, when people buy a home, they might want to change from the main contract type of the general medical insurance products to the special accessory contract type of the life insurance product. This is a finding that the insurance company mainly dealing with the life insurance products should recommend the reexamining of the purchase condition of the insurance products to the existing and potential customers when they buy a home.

7 Conclusions

In this study, we proposed a new methodology of modeling the agent's behavior in ABSS using Bayesian network based on the questionnaire survey. The main contribution of the proposed method is that it enables us to simultaneously perform the construction of the agent's behavior model and the estimation of the internal parameters within the model. Taking medical insurance market as an actual case, we built the agent-based model including the agent behavior model that reflected the realities of consumer's behavior. We executed the ABSS using the resultant model and then verified the effectiveness of the proposed methodology by investigating the results derived from the simulation. As a result of the analysis under the 12 different scenarios, it was found that the effectiveness of the sales promotion of marketing, especially the sales activity and the media advertisement, is different depending on the kind of the medical insurance. Also it was found that the life stage of people affects the choice of the medical insurance product.

Acknowledgment This work was supported by the Grant-in-Aid for Scientific Research (B) from the Japan Society for the Promotion of Science (JSPS), JSPS KAKENHI Grant Number 26282087.

References

1. Wang, F. Y., Carley, K. M., Zeng, D., & Mao, W. (2007). Social computing: From social informatics to social intelligence. *IEEE Intelligent Systems, 22*, 79–83.
2. Macal, C. M., & North, M. J. (2010). Tutorial on agent-based modeling and simulation. *Journal of simulation, 4*(3), 151–162.
3. Axelrod, R., & Tesfatsion, L. (2006). A guide for newcomers to agent-based modeling in the social sciences. In *Handbook of computational economics, Vol. 2: Agent-based computational economics, Hand books in economic series* (pp. 1647–1658). Amsterdam: North-Holland.
4. Ohori, K., Iida, M., & Takahashi, S. (2013). Virtual grounding for facsimile model construction where real data is not available. *SICE Journal of Control, Measurement, and System Integration, 6*(2), 108–116.
5. Heckerman, D., Geiger, D., & Chickering, D. (1995). Learning Bayesian networks: The combination of knowledge and statistical data. *Machine Learning, 20*, 197–243.
6. Jensen, F. V. (2001). *Bayesian networks and decision graphs*. New York: Springer.
7. Holmes, D. E., & Jain, L. C. (Eds.). (2008). *Innovations in Bayesian networks: Theory and applications, Studies in computational intelligence, 156*. Berlin: Springer.
8. Chen, X., Anantha, G., & Lin, X. (2008). Improving Bayesian network structure learning with mutual information-based node ordering in the K2 algorithm. *IEEE Transactions on Knowledge and Data Engineering, 20*, 628–640.
9. Weiss, Y. (2000). Correctness of local probability propagation in graphical models with loops. *Neural Computation, 12*, 1–41.
10. Mooij, J., & Kappen, H. (2007). Sufficient conditions for convergence of the sum–product algorithm. *IEEE Transactions on Information Theory, 53*, 4422–4437.
11. Kocabas, V., & Dragicevic, S. (2012). Bayesian networks and agent-based modeling approach for urban land-use and population density change: A BNAS model. *Journal of Geographical Systems, 15–4*, 403/426.
12. Shen, Y., Liu, S., Fang, Z., & Hu, M. (2012). Modeling and simulation of stranded passengers' transferring decision-making on the basis of herd behaviors. *Kybernetes, 41*(7/8), 963–976.
13. Miyazaki, M., Ishino, Y., Takahashi, S. (in press). Effects of word-of-mouth communication on product diffusion: A case of medical insurance product, In *Post-proceedings of the AESCS international workshop 2013, agent-based approaches in economic and social complex systems VIII*, Springer.
14. Kuribayashi, A. (2006). *Impact of life insurance advertisement to consumer's awareness and behavior*. Report of Nissay Basic Res. Center, REPORT 2006.1 (In Japanese)
15. Tanaka, T. (2009). Consideration concerning sales representatives of life insurance products. *Journal of Life Insurance, 169*, Japan Institute of Life Insurance (In Japanese).
16. Ishino, Y. (2014, March 5–7). *Consumer survey data analysis to model internal state of an agent*. Proceedings of 5th symposium of technical committee on social systems, pp. 197–200, Okinawa, Japan (In Japanese).
17. Kuribayashi, A. (2008). *Possibility of word-of-mouth in marketing of life insurance products*. Report of Nissay Basic Research Center, REPORT 2008.4 (In Japanese)
18. Bollobás, B. (2001). *Random Graphs* (2nd ed.). Cambridge: Cambridge University Press.

Coping with Bullying in the Classroom Through Agent-Based Modeling

Chaiwat Thawiworadilok, Mohsen Jafari Songhori, and Takao Terano

Abstract In the recent year, observation of *Ijime* among Japanese students has been a growing concern. To understand the effect of this behavior in a detail level, we build an agent-based model and conduct a set of experiments. In the model, the main interactions occur between victim and bully as prisoner dilemma game. The results indicate that students can attend higher social standing by being not cooperative. As such, if one is a victim, it is recommended not to comply with the bully. In the case of being bullied, one needs to take the role seriously. By encouraging victim to be aggressive toward bully, the effect of *Ijime* can be alleviated.

Keywords Iterated prisoner's dilemma • Ijime • Bully • Agent-based modeling

1 Introduction

In the recent year, 「いじめ」 or *Ijime* in *Romaji* has been one of the concerns among Japanese students at school. Not only face-to-face form, but it is also in the online environment. It is one of the major factors that leads to suicide in many cases [1]. Besides resulting in suicide, there were many reports in the case of social withdrawal, hikikomori (ひきこもり). One of the leading causes of *Ijime* comes from the Japanese culture of conformity. As such, the strange or weak student is more likely to become the victim of such a case.

According to [2]), *Ijime* appears to be similar to western girls' bullying behavior which focuses on mental aggression. However, *Ijime* structure differs from western

C. Thawiworadilok (✉) • T. Terano
Tokyo Institute of Technology, Interdisciplinary Graduate school of Science and Engineering, Computational Intelligence and system Science J2 Bldg., Room 1704, 4259 Nagatsuta-cho, Midori-ku, Yokohama, Kanagawa 226-8502, Japan
e-mail: teerachait@hotmail.com

M.J. Songhori
Tokyo Institute of Technology, Interdisciplinary Graduate school of Science and Engineering, Computational Intelligence and system Science J2 Bldg., Room 1704, 4259 Nagatsuta-cho, Midori-ku, Yokohama, Kanagawa 226-8502, Japan

JSPS Research Fellow, Tokyo, Japan

U.S. Putro et al. (eds.), *Agent-Based Approaches in Economics and Social Complex Systems IX*, Agent-Based Social Systems 15, DOI 10.1007/978-981-10-3662-0_5

bullying as the role of bully is always changing. In other words, the best and most obedient student in the class can become the bully. The role of students in the class can be categorized into four types, bully, supporter, victim, and bystander [3]. Bystander main action is to avoid getting involved with *Ijime*, while supporters are the ones who enjoy *Ijime* along with the bully. One of literatures [3] suggests that the bystander behavior encourages bullying as it allows atmosphere of permissiveness.

Agent-based modeling is one of a computational model which uses actions and interactions among agents to understand a system as a whole [4, 5]. The main feature of agent-based modeling is to use microlevel architecture/interactions to understand macrolevel emergence. Due to the difficulties in real classroom experiment, agent-based modeling is used to simulate the classroom situation. It provides an alternative to the real experiment that can be dangerous to implement or difficult to be observed efficiently. Also, agent-based modeling is able to overcome the problem of observing a phenomenon for a long time period.

We build a model for *Ijime* that its main parts are bully and victim. In addition, prisoner's dilemma (here after PD) game is used to represent the interaction between the two. The similarity to PD game comes from the fact that either party can choose to defect or cooperate. To a large extent, PD game represents cases where rational agents choose not to be cooperative even when the situation seems to encourage both parties to cooperate [6–9]. Other researchers [10] also use PD game to simulate the *Ijime* situation on bystander interactions. This paper focuses more on the victim and bully interactions as well as including supporter and bystander roles inside the model.

The rest of the paper is structured as the following: "Model" is a part which described the *Ijime* model in detail. Then in the "Experiment" part, results and graphs of the simulation are shown with an explanation. Followed by "Discussion and Conclusion" part, the emergent pattern including its interpretation is explained in this section. "Extension" part is about possible work which can be done in the future.

2 Model

The model description follows the overview, design concepts, and detail protocol for describing agent-based model [11].

2.1 Purpose

The agent-based modeling is designed in order to investigate *Ijime* behavior. Through it, we aim to find any useful insight regarding the matter. Consequently, we hope that with this model there could be alternatives to deal with *Ijime* in an efficient way. By doing so, the problems from social withdrawal and suicide can be alleviated.

2.2 Entities, State Variables, and Scales

In this model, there is only one entity which is students in a class. At each time step, each agent is assigned one of the four roles, bully, victim, bystander, and supporter. The state variables and description are shown in Table 1.

In this model, we have a number of agents. That is shown by n. These agents are shown in $\mathrm{SA} = \{\mathrm{i} | \mathrm{i} = 1, 2, \ldots, n\}$. Also, these agents belong to a particular group. So there are n_g groups as $\mathrm{SG} = \{\mathrm{g} | \mathrm{g} = 1, 2, \ldots, n_g\}$. At time t, $G^t_{i,g}$ indicates that agent i belongs to group g. The score of agent i at time t is represented by s^t_i. Let S^{t-1}_i refers to the cumulative score at time t; thus, $S^{t-1}_i = \sum_{t'=1}^{t-1} s^{t'}_i$. In addition, we use the number of times agent i got selected in each role to determine agent characteristic. The roles are nv^{t-1}_i for victim, nb^{t-1}_i for bully, ns^{t-1}_i for supporter, and nby^{t-1}_i for bystander.

Furthermore, we control parameters according to the following:

- Number of members in one group: {5, 10}
- Ratio between supporters and bystanders: [0.25, 0.75]
- Number of people in one class: {30} [reference 12]
- Number of iterated games: {500, 1000}
- Probability of becoming helper: [0.025, 0.075]

Regarding the probability of becoming a bully, many of the literatures suggest that the role of bully is rotated among friends in the same group [13–15]. As a result, p^{bully}_i is assumed to be uniformly distributed. The probability of becoming a victim is calculated as follows where k refers to the agent in the same group with agent i:

$$p^{\text{vic}}_i = 1 - \frac{S^{t-1}_i}{\sum_{k=1}^{n} S^{t-1}_k} \quad ;\text{for all agent} \tag{1}$$

Table 1 State variable and description

State	Description	Variable name in the model
Characteristic vector	Cumulative number of being chosen as victim, bully, supporter, and bystander at time $t-1$	$< nv^{t-1}_1, nb^{t-1}_1, ns^{t-1}_1, nby^{t-1}_1 >$
Role probability	Probability of being chosen as victim, bully, bystander, and supporter, respectively	$p^{\text{vic}}_1, p^{\text{bully}}_1, p^{\text{bys}}_1, p^{\text{sup}}_1$
Score	Score at time t	s^t_i
Cumulative score	Cumulative score of an agent at time $t-1$	S^{t-1}_i
Change group probability	Probability of changing to group g	$p^{ch}_{1,G^t_i,g'}$

There exists dynamics not only within the group but also between groups. Hence, students may change their groups at the different times. In this case, changing group probability is calculated from victim probability and similarity of characteristic vector as follows:

$$p^{ch}_{i,G^t_i,g^\cdot} = w * p^{\text{vic}}_i + (1-w)$$
$$\left(\frac{\sum_{j\in g_\cdot}\left(\left|nv^{t-1}_i - nv^{t-1}_j\right| + \left|nb^{t-1}_i - nb^{t-1}_j\right| + \left|ns^{t-1}_i - ns^{t-1}_j\right| + \left|nby^{t-1}_i - nby^{t-1}_j\right|\right)}{4 * |g_\cdot| * t - 1}\right) \quad \text{; for all agents} \tag{2}$$

Let sv^t_i and sb^t_j indicate the score of victim and bully agents from the PD game, respectively (if agent i and j are victim and bully in this order). This definition involves time t as the game is repeated over time.

For supporter score, it is calculated as the following equation:

$$s^t_i = \left(\frac{\text{number of supporter}}{n-2}\right) * \left(sb^t_i + 1\right) \quad \text{; agent } i \text{ is the supporter} \tag{3}$$

The logic behind the calculation is that the supporters are the ones agreeing with the bully behavior. The score that supporters receive is lower than the bully score as supporters don't cause the event to happen directly. Furthermore, the bully also receives an additional score as follows:

$$s^t_i = sb^t_i + \left(\frac{\text{number of supporter}}{n-2}\right)\left(sb^t_i + 1\right) \quad \text{; agent } i \text{ is the bully} \tag{4}$$

Other than the former types, when bystander agents do nothing or give help to the victim (and are not seen), their scores are assumed to be the same. As such, we decided to use multiplier 1 as the situation where the agent is not seen is similar to the situation where the bully and victim are fighting. The bystander score is calculated as follows:

$$s^t_i = \left(\frac{\text{number of supporter}}{n-2}\right) * 1 \quad \text{; agent } i \text{ is the bystander} \tag{5}$$

Generally, bystanders tend to avoid any conflict with *Ijime* incident. As a consequence, a bystander can afford a better standing than the victim. On the contrary, the seriousness of *Ijime* also depends on the number of supporters participating. When bystanders got seen by their peers, they will get score as follows:

$$s^t_i = -\left(1 + \frac{\text{number of supporter} + 1}{n-2}\right) * 5 \quad \text{; agent } i \text{ is the bystander} \tag{6}$$

The probability of being seen is described as:

$$p_i^{\text{seen}} = \frac{\text{number of supporter} + 1}{n - 2} \quad \text{; agent } i \text{ is the bystander} \tag{7}$$

In the situation where there are helps from bystanders, the victim receives additional score from helper as follows:

$$s_i^t = sv_i^t + \left(\frac{\text{number of helper}}{n - 2}\right) * 1 \quad \text{; agent } i \text{ is the victim} \tag{8}$$

After the score of the current game is finalized, the next step is to finalize the current score. The calculation is as follows:

$$S_i^t = S_i^{t-1} + \left(\frac{\text{current game score}}{\text{Max}\left\{ s_{\text{sup}}, s_{\text{bully}}, s_{\text{bys}}, s_{\text{vic}} \right\}}\right) \quad \text{; for all agents} \tag{9}$$

2.3 *Process and Overview*

At each time step, each agent is assigned to a role according to agent characteristic. The following pseudocode shows how each agent operates in one cycle (Fig. 1):

Fig. 1 Pseudocode

```
Pseudo code (within group interaction at time step t)
For each group
      For each agent in group
            Bully role is selected randomly
                  Victim role is selected (1)
                  Supporter and bystander are selected
      Let bully and victim PD game (PD table) (PD table)
      Supporter update score (3)
      Bystander update score (5)(6)(7)
      Bully update score (4)
      Victim update score (8)
     For each agent in group
                  Normalize score (9)

            Update the P_Vic P_goout Score_old (1), (2)

For every agents

      Set P_Vic P_goout Score_old (1), (2)

      Decide whether to change group or not
Increase time step to t+1
End of time step
```

Table 2 PD game for bully and victim

Bully	Victim	
	Cooperate	Defect
Cooperate	R = 3,3	T = 5,S = 0
Defect	T = 5, S = 0	P = 1,1

2.4 *Design Concepts*

The interaction of victim and bully is modeled as the PD game. In the *Ijime* situation, both bully and victim can choose either to cooperate or defect. The following table shows the payoff for each pair of interactions Table 2.

From the perspective of *Ijime*, the four cases of the game can be described as follows:

1. The case where bully and victim cooperate refers to the situation where bully only makes little joke on victim and victim does not take it into mind.
2. When bully defects and the victim cooperates, it refers to the situation where the bully aims to bully the victim and the victim does not fight back.
3. The case where bully cooperates and victim defects refers to the situation where bully jokes with the victim, but the victim takes it seriously and fights back the bully.
4. Finally, in the case where both of them defect, the victim and the bully are fighting against each other.

Agents can revise their strategies for the cases wherein they become either bully or victim. By using survival of the fittest concept [3], the top 60% strategies will be redistributed among the members of class every 100 time steps. This indicates the learning of the agent within group as the successful strategies thrive. The game continues according to the number of iterated game defined in the parameter's range table. Figure 2 shows how the game is played out.

2.5 *Initialization*

At the beginning of each simulation, every agent is assigned randomly to three prior histories of actions, either defection or cooperation. This is to make strategies which require history to be able to be usable at the first time step. Regarding the victim probability for each agent, the value is set as $p^0_{\text{vic}} = \left(\frac{1}{\text{number of people in group}}\right)$ according the roulette wheel algorithm [16]. Furthermore, each strategy is distributed evenly at the first time step.

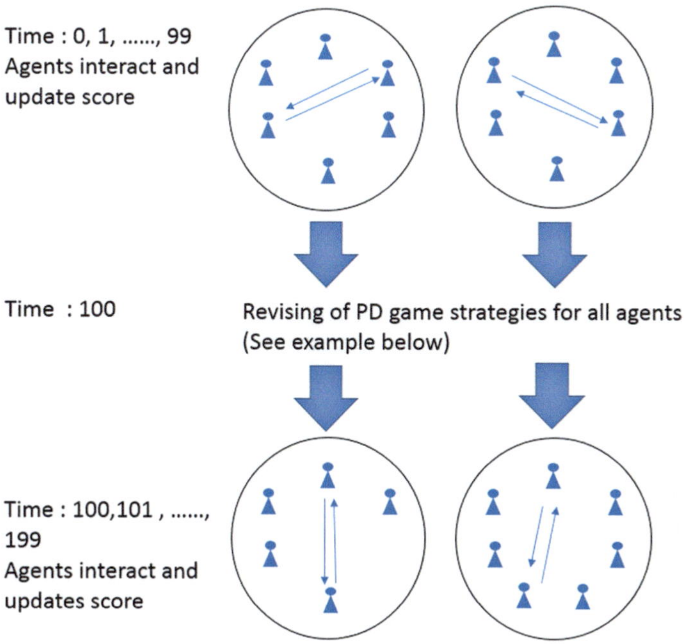

Fig. 2 Game played from the first round until 200th round

3 Experiment

For the number of students in the class, the number varies depending on the type of the school, according to the literature [12]. Table 3 indicates the strategies which are used by victims and bullies in the simulation.

Nevertheless, the average between both cases is used. At the base case for this experiment, each group has no interactions among themselves apart from agents moving between groups. Through all the parameter's range, there can be 16 possible scenarios.

Agents play game according to the allocated strategies. The victim and bully histories are updated, while supporters' and bystanders' histories stay the same.

4 Results

For each run, the simulation is generated using the parameters defined in the model section. To control randomness of the generated values and get insights, thousand simulations were generated in order to find the average value. As most of the results tend to have similar trend, the following setting is selected: (I) five members per group, (II) ratio of supporter to bystander ratio = 0.25, (III) probability of bystander becoming helper =0.075, and (IV) $T_{\text{sim}} = 1000$ ticks.

Table 3 Strategies for bully and victim

Strategy	Description
Always cooperate	Always play cooperation
Always defect	Always play defection
Tit for tat	Starting the game by cooperation, then using the last play from other players
Tit for 2 Tat	Starting the game by cooperation, then using the last two plays from other players
Comply to the strong bully (only for victim)	The more the score differs, the higher the probability to become cooperative. If victim score is higher, then it will not cooperate
Bully the weak victim (only for bully)	The more the score differs, the higher the probability to become defective. If victim score is higher, then it will not defect
Cooperate with similar agent	The less the score differs, the higher the probability to become cooperative

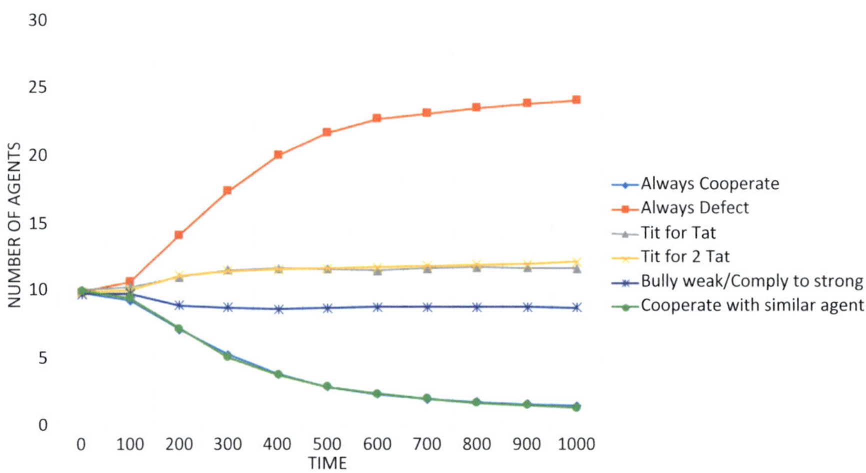

Fig. 3 Average number of agent for each strategy

The following plot (Fig. 3) shows the average number of agents for each strategy. According to this figure, the most popular strategy is "always defect." On the other hand, the strategies which are fading out through time are "always cooperative" and "cooperate with similar agents." Even though the graph starts with the same number of agents, the strategy with higher score is more likely to pass on the strategy to the next generation.

Figure 4 illustrates the average score of agents for each strategy. It also shows that "always defect" strategy has the highest average score throughout the period. In addition, "always cooperate" and "similar agent" strategies produce the lowest score range. The observed behavior in Fig. 4 is consistent with what has shown in Fig. 3.

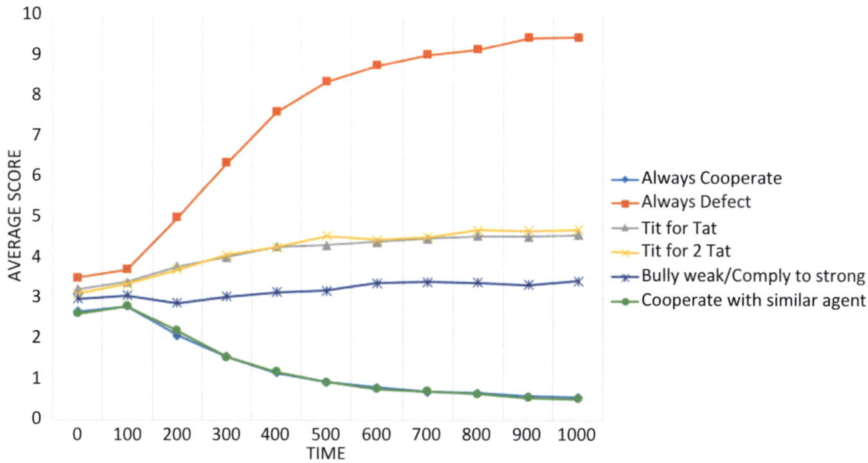

Fig. 4 Average score of each strategy

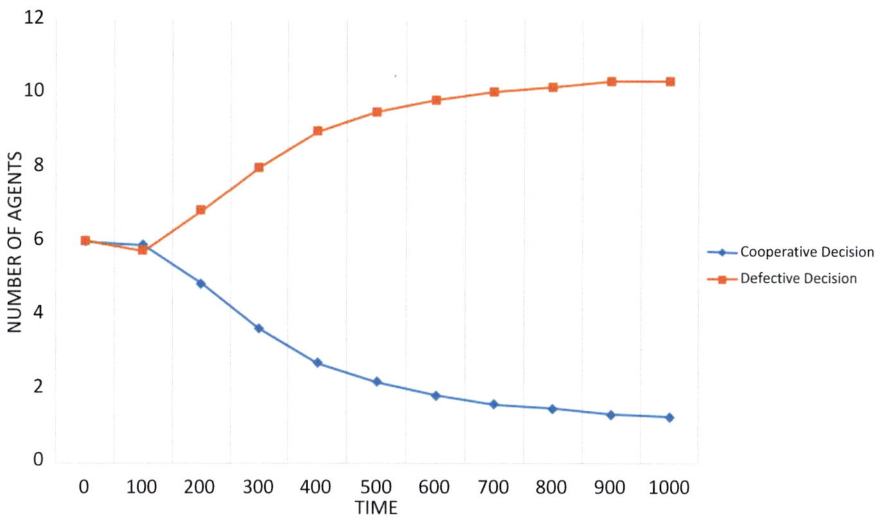

Fig. 5 Average of cooperative and defective decision at each time step

Figure 5 illustrates the average number of cooperative/defective decision at each time step. This figure shows that defective becomes more popular over time.

5 Discussion and Conclusion

We have developed model for Ijime behavior. By using PD game, the interactions in classroom were simulated. The emergence pattern can be observed through the presented graph. The results from the experiments are also robust throughout the parameter's range. With all 16 possible combinations of the parameter's range, the generated simulation results tend to have similar trend.

- It appears that cooperation does not work well in this environment. This reflects current *Ijime situation* in Japan where victims are forced to cooperate with the group. Generally, the chosen victims are told to put up with the bully in order to keep the flow of the class. The situation grows worse for the chosen victim as he/she is likely to be chosen as victim again. With this repeated cycles, it can lead victim to serious conditions.
- Alternatively, the chosen victim can choose to disrupt the class harmony by standing up and expressing his/her feeling. The experiment suggests that the victim can get a higher social position in a classroom if he/she raises up the dissatisfaction. As such, by not letting anyone bully you for free, one can reduce the effects of *Ijime* to a certain level.
- To become a bully, any student in a group has two options: by volunteering themselves or by group pressure. From the experiments, it can be observed that the chosen bully also loses its social value if they keep on cooperating with the chosen victim. When the bully jokes with the victim, there is a probability that the chosen victim may try to retaliate in the future.
- In order to keep up their social value in the class, it is recommended for the one who plays bully roles to be more serious about his/her task as bully. By doing so, the chosen bully can keep up his/her social value in the classroom.

In summary, using an agent-based simulation, we have gained better understanding on how to deal with *Ijime*. By encouraging the victim to be brave against the bully, we can reduce the effect of *Ijime* on the victim. Furthermore, we can also help the unintentional bully to get along well with the class by encouraging them to be more conform to his/her group. This means that the chosen bully must take his/her role seriously.

Similar to any other study, we have made some assumptions/limitations in this paper. For future work, we plan to compare our model with empirical observation of effectiveness of the recommended strategies. In addition, improving the model by using mathematical model is also being considered. Furthermore, using another kind of game theory to build similar agent-based modeling to validate the result of current model is another venue.

References

1. Takashi, N., et al. (2005). Bullying and ijime in Japanese schools: A sociocultural perspective. In *Violence in Schools* (pp. 169–190). New York: Springer.
2. Taki, Mitsuru. (2003, May 21–25). 'Ijime bullying': Characteristic, causality and intervention, Oxford-Kobe Seminars.
3. Yohji, M. (1996). Bullying as a contemporary behaviour problem in the context of increasing 'societal privatization' in Japan. *Prospects, 26*(2), 311–329.
4. Kobayashi, T., et al. (2013). Is there innovation or deviation?-Analyzing emergent organizational behaviors through an agent based model and a case design. In eKNOW 2013, The fifth international conference on information, process, and knowledge management, pp. 166–171.
5. Takahashi, S., et al. (2014). How can start-up business firms keep the motivations of employees?: Analyzing organizational management strategies through an agent based model. In Proceedings of the 6th international conference on information, process, and knowledge management (eKNOW) IARIA, 2014, pp. 1–6.
6. Axelrod, R. (1986). An evolutionary approach to norms. *American Political Science Review, 80*(04), 1095–1111.
7. WU, J., & Axelrod, R. (1995). How to cope with noise in the iterated prisoner's dilemma. *Journal of Conflict Resolution, 39*, 183–189.
8. Axelrod, R. (1987). The evolution of strategies in the iterated prisoner's dilemma. In *The dynamics of norms* (pp. 1–16). Cambridge: Cambridge University Press.
9. Axelrod, R., & Hamilton, W. D. (1981). The evolution of cooperation. *Science, 211*(4489), 1390–1396.
10. 矢野翔太; 近匡; 小柳文子. 囚人のジレンマを用いたいじめ発生メカニズムの解析と対策, (2013).
11. Grimm, V., et al. (2006). A standard protocol for describing individual-based and agent-based models. *Ecological Modelling, 198*(1), 115–126.
12. OECD. (2010). How many students are in each classroom? In *Highlights from education at a glance 2010*. Paris: OECD Publishing.
13. Taki, Mitsuru. 2001. Japanese school bullying: ijime. Ponencia presentada en la jornada "Comprendiendo y preveyendo el acoso escolar: una perspectiva internacional", celebrada el, 19.
14. Peaslee, E. 2011. *Adolescent suicide in Japan: The fatal effects of Ijime.*
15. Yoneyama, Shoko. 2008. *The era of bullying: Japan under Neoliberalism.*
16. Birk, A. (2001). Learning to trust. In *Trust in cyber-societies* (pp. 133–144). Berlin/Heidelberg: Springer.

How to Attract Customers to Your Website with Word-of-Mouth Communication in Social Media

Jieliang Zhou, Takashi Yamada, and Takao Terano

Abstract The social environment of a person has a high influence of his/her consumer behavior. Social networks transfer these social environments to the online world and enable a targeted influence on the customers purchasing behavior by word of mouth from their social contacts. In this paper, we present an agent-based simulation model that enables computational experiments of online media's management process using the word-of-mouth communication in social media. By comparing the number of customers' visits of two viral media in competitive setting, we confirm that following the market trend to make the contents and deliver them to the user who has more friends or relationships in social media is the effective way to gain customer's visit from social media.

Keywords Online media • Social network service • Word-of-mouth communication

1 Introduction

Rapid growth of the social media has not only changed the way people communicate and gather on the web but also affected the way of content discovery and navigation in a big way. According to "Reuters institute digital news report 2014" [1], more and more users, especially the young users, have become to choose social media as their gateway to make use of the online media contents. Meanwhile, the social media marketing which centers on creating content that attracts attention and encourages readers to share it across their social networks is grabbing a great deal of attention.

J. Zhou (✉) • T. Terano
Tokyo Institute of Technology, Kanagawa, Japan
e-mail: jeremysyu.liang@gmail.com

T. Yamada
Yamaguchi University, Yamaguchi, Japan

U.S. Putro et al. (eds.), *Agent-Based Approaches in Economics and Social Complex Systems IX*, Agent-Based Social Systems 15, DOI 10.1007/978-981-10-3662-0_6

"Viral media" is a new media style which employs the social media marketing to receive the majority of its traffic by creating content that is shared on social media websites. As a consequence, how to make the contents popular in social media is the most significant issue that the marketers there should take into consideration. The famous viral media site "BuzzFeed" reaches more than 130 million unique visitors per month which almost surpass most of the traditional online website [2]. On the other hand, compared to the traditional online website, because of the easiness to begin with a new one, varieties of viral media have been mushrooming, and the fierce battles among the viral media sites have been observed. For this reason, it is not easy for the viral media sites to survive under such business environment.

In this study, we construct an agent-based simulation model to reproduce the real-world management process of viral media to explore the feature of the viral media. By constructing two viral media in the competitive setting, we will indicate the effective contents of making strategy and targeting strategy to gain more customers from social media.

2 Related Work

To model and manage the word-of-mouth communication process effectively, Bampo et al. [3] proposed a decomposition approach of word-of-mouth activity consisting of three main aspects: (1) the particular structure of the social network, (2) the behavioral characteristics of its constituting members, and (3) the seeding strategy to initiate the viral process. We conceptualize our model by this approach to study the dynamics of viral media management process. However, our work differs from it by considering the change of the promotion contents, which can assist the online media marketer's to decide how to update their contents in diffusion process. We also analyze the promotion strategy in the model ground with the reality.

3 Model Outline

We assume that there is a competitive environment with two online media companies utilizing the word-of-mouth communication to attract customers from SNS and wherein the user agents consume the contents the companies create based on their rules. Every user has an individual circle of friends consisting of other user agents. Figure 1 shows the outline of the proposed model. In our model, we have three kinds of agents, the news pool, the online media agent, and the consumer agent.

In every time step, the consumer users read and share the contents. After every several time steps, the news pool generates several media contents, and the online media agents select the new contents as the delivery contents and change the target SNS users. The behavioral rules for users and media agent are shown in Fig. 2.

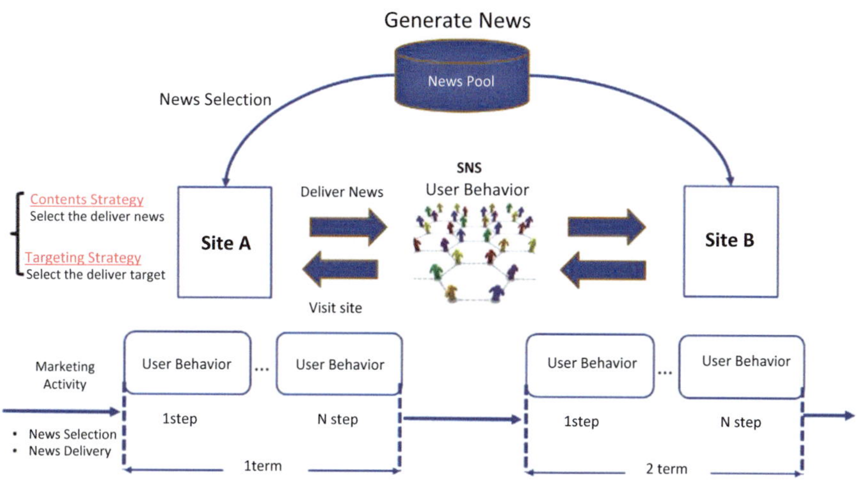

Fig. 1 Outline of the proposed model

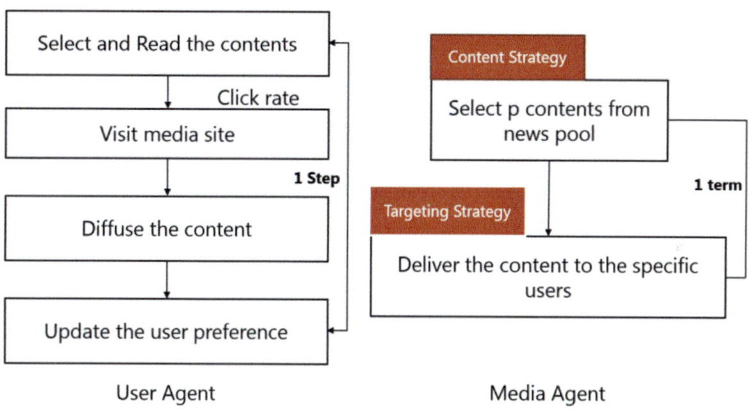

Fig. 2 Behavioral rules for each agent type

Media agents will (1) select the contents from the generated contents from news pool based on their content strategy and (2) transmit the contents to the target consumers based on their targeting strategy. Consumer agents select one content which can maximize their own utility from the received contents and then read and visit the media site according to the static click rate. Then, they diffuse the contents and update the personal preference.

The previous process of interaction between media agent and consumer agents is repeated over time. In addition, at the end of each time interval (i.e., terms), the media agents revise their content strategy.

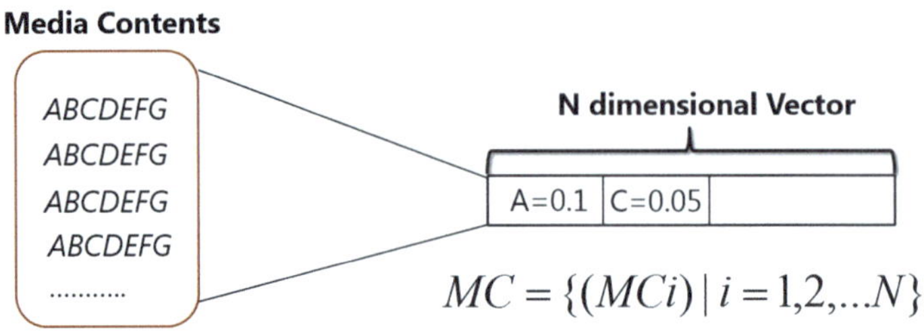

Fig. 3 Image of online media contents

3.1 News Pool

The news pool agent will generate M contents in every term (several steps). We use an N-dimensional vector to represent the online media content that is used by media agent. Each element of this vector indicates a keyword by which a user can reach the media content. In other words, each of the N elements can be conceptualized as one feature of the media content. The elements of the N-dimensional vector include not only the keywords but also a quantified weight. This weight shows the percentage of the media content in which the keyword is repeated or its synonym appears. Figure 3 shows the image of the online media contents in our model.

3.2 Consumer Agents

The same N-dimensional vector used for media content is considered to describe each user's preference. The segmentation set of keywords is defined as $UP = \{UP_i | i = 1, 2 \ldots N\}$. UP_i is the feature value of the user attribute i.

The consumer agents will receive the contents, adopt the contents, and diffuse the contents according to their own activity rules: the quality of the contents and the social impact from the adjacent users in social network.

We use utility to represent the extent of acceptance of the received content for each user. By summing up each agent's content utility and social utility, the weighted linear sum is determined as follows:

$$Ut_i(Mc) = W_i * Uc_i(Mc) + (1 - W_i) * Us_i(Mc) \tag{1}$$

where $Uc_i(Mc)$ is the content utility for the received content Mc, $Us_i(Mc)$ is the social utility, and W_i is the weight for content utility.

We use the similarity between user preference vector and media content vector to define the content utility:

$$Uc_i(Mc) = \cos(Mc, Up_i) \quad U_c \in (0, 1) \tag{2}$$

Social utility can be defined as the following equation based on the information diffusion model proposed by Rand and Rust [4]:

$$U_s(Mc) = \frac{n_t(Mc)}{n} \quad U_s \in (0, 1) \tag{3}$$

where $n_t(Mc)$ is the number of content Mc received from adjacent node set S and n is the number of adjacent nodes.

3.2.1 Content Selection Rule

The consumer agent can only choose one content from received contents in every time step which means the agent who received several contents will select one content based on the following roulette wheel selection algorithm:

$$P_k = \frac{Ut_i\left(Mc^k\right)}{\sum_{k=1}^{Np} Ut_i\left(Mc^k\right)} Mc^k \in \text{ } N\text{pool}^i \tag{4}$$

where $Ut_i(Mc^k)$ denotes the utility value for the kth media contents received in current time which is stored in the media content's storage $Npool$ for user i. Np is the number of received contents.

3.2.2 Content Diffusion

Consumer agents in the social network can diffuse the contents they receive from viral media based on specific rules. Here, we define the information diffusion cost, Ci, to control the user's diffusion probability based on the information diffusion model of Hildebrand et al. [5]. The consumer will diffuse the viral media content only when their utility for the content is bigger than their diffusion cost. The diffusion rule is written as follows:

$$\delta = \begin{cases} \text{Diffuse, } \textbf{if } Ut_i(Mc) - C_i > 0 \\ \hline \text{Do not diffuse, } \textbf{otherwise} \end{cases} \tag{5}$$

3.2.3 Change of the User Preference

We consider the effect of user's friendship and external environment to the change of the user preference. We define the effect of user's social relationship as the local effect and update the user's preference as follows:

$$Up_i^k(t) = Up_i^k(t) + \sum\nolimits_{j \in s} \alpha_t \left(Mc_j^{kd}(t) - Up_j^{kd}(t)\right) \tag{6}$$

where UP_i^k (t) is the feature value for user i's feature k, α_t is the value of α in current step t, MC_j^{kd} is the feature value kd of media contents received from user j in current time step, UP_i^{kd} is the feature value kd of user i's user preference in current time step, and S is the set of users which has link with user i in social media.

Furthermore, in order to represent the process that the consumers are attracted by new information and get bored with the old one with the lapse of time, we define the local effect coefficient α as the following exponential decay function:

$$\alpha_t = \alpha_0 e^{-\gamma t} \tag{7}$$

where α_t is the value of α at step t, α_0 is the initial value of α, t is the elapsed step, and γ is the attenuation coefficient and set as 1 in this model.

Similarly, to represent the global environment like the mass media and trend of the society, we use the following equation to update the user agent's preference:

$$UP_i^k(t) = Up_{i\,i}^k + \beta_t \left(Mc_{\text{global}}^k - UP_i^k \right) \tag{8}$$

where $UP_i^k(t)$ is the feature value for user i's feature k, β_t is the value of β in current step t, and MC_{global}^k is the feature value k of global media contents. In each time step, the global media content MC_{global} is delivered to every user in social media.

3.3 *Media Agents*

Media agents can change their management strategy regularly in order to gain more visitors to their website. Here, we explain the targeting strategy and content marketing strategy as their management strategy.

3.3.1 Content Marketing Strategy

Media agents can change and update the delivery contents in each term to attract more visitors to the website. The following strategies are identified and developed in our interviews and discussions with online media that operates in Japan:

1. Learning Strategy

This strategy will select the contents from the news pool according to the number of customers attracted by the contents delivered by own company in previous term. This is comparable to the A/B test which is widely used in real-world website marketing. So, the media agent using this strategy will choose the contents which are most similar with the most attractive contents delivered by own company in previous term according to the following equation:

$$P_{\mathrm{del}}^{k}(t) = \frac{\cos\left(Mc^{\mathrm{Obest}}(\mathrm{t}-1), Mc^{k}(t)\right)}{\sum_{i=1}^{M}\cos\left(Mc^{\mathrm{Obest}}(t-1), Mc^{i}(t)\right)} \tag{9}$$

where $P_{del}^{k}(t)$ is the probability to select content k as the delivery contents from news pool, Mc^{Obest} $(t-1)$ is the contents achieved most of the customers from social media in own company, $Mc^{i}(t)$ is the ith contents in news pool, and M is the number of contents in news pool.

2. Trend Strategy

This strategy will select the contents from the news pool according to the number of customers attracted by the contents delivered by both own and opponent company in the preceding term. This strategy can reproduce the imitation of other companies in the real world. So, the media agent using this strategy will choose the contents which are most similar with the most attractive contents delivered in market in previous term according to the following equation:

$$P_{\mathrm{del}}^{k}(t) = \frac{\cos\left(Mc^{M\mathrm{best}}(\mathrm{t}-1), Mc^{k}(t)\right)}{\sum_{i=1}^{M}\cos\left(Mc^{M\mathrm{best}}(t-1), Mc^{i}(t)\right)} \tag{10}$$

where $P_{\mathrm{del}}^{k}(t)$ is the probability to select content k as the delivery contents from news pool, $Mc^{M\mathrm{best}}$ $(t-1)$ is the contents achieved most of the customers from social media in both own and opponent company, $Mc^{i}(t)$ is the ith contents in news pool, and M is the number of contents in news pool.

3.3.2 Targeting Strategy

This strategy encourages a certain number of user agents to immediately adopt their media contents. However, the most challenging problem for marketing manager to use this social media platform successfully is to have consideration about the privacy of users. Since the entire network structure is not available for the advertisers, they are forced to rely on the third-party information about the targeting consumers. In order to simulate this feature, we define a targeting strategy considering the network statistic characteristics instead of a whole network to find the most effective targeting strategy for the online media site's promotion under different situations. The possible strategies are as follows:

1. Degree Strategy

This strategy is to deliver the contents selected from the news pool to the target users according to the degree of the network. Higher-degree nodes influence more neighbors, directly encouraging more adoption. This strategy is comparable to marketing strategy choosing the user who has more friends to deliver the contents in the real world. So the number of neighbors of the target node normalized by the maximum possible value described as the equation below is calculated and delivers the contents to the users who have higher $w_d(i)$ value:

$$w_d(i) = \frac{\text{degree}_{(i)}}{\max\,(\text{degree})} \tag{11}$$

2. Clustering Coefficient Strategy

This strategy is to deliver the contents selected from the news pool to the target users according to the clustering coefficient of the network. The lower the clustering coefficient of a node, the less overlap there is among its neighbors, encouraging wider adoption more quickly. So the 1.0 minus the fraction of neighbors of the node whose neighbors are also neighbors of the target node is calculated according to the following equation, and the media agent using this strategy will deliver the contents to the users who have higher $w_c(i)$ value:

$$w_c(i) = 1.0 - \frac{cc(i)}{\max(cc)} \tag{12}$$

4 Computational Experiments

In this study, we assume there are two viral media companies, Viral A and Viral B, in the competitive environment and attract the visitors to their website based on their management strategy. In this experiment, we set the content strategy and targeting strategy for each media agent according to the strategy described above and compare the number of attracted customers of the two companies to find the effective content selection and content delivery strategy for the viral media site to attract the customers. We use an improved version of Barabasi-Albert network [6] called DMS model which has the scald-free feature and small-world feature that is widely observed in social network as a static network to represent the real-world SNS network. The pair of the content strategy and targeting strategy is shown in Table 1.

The parameters setting for simulation is shown in Table 2.

Table 1 The pair of the strategy

	Degree strategy	Clustering coefficient strategy
Learning strategy	(1)	(3)
Trend strategy	(2)	(4)

Table 2 Simulation parameter

Parameter	Value
Ci: content diffusion cost	Uniform(0.25,0.5)
Wi: weight of the content utility	Uniform(0,1)
α: local effect of the change of the user preference	Uniform(0,1)
β: global effect of the change of the user preference	Uniform(0,1)
number_population: number of the social population	10,000
Term_step: number of steps in one term	24
M: number of news generated in each step	10
P: number of contents selected from news pool in each tern	3
q: click rate for the reading contents	0.1
γ: power index for the network	2
Step: simulation step	1000

Table 3 Competition result (100 trails)

	①	②	③	④
①		79	84	23
②	21		44	11
③	16	56		18
④	77	89	82	

The value of upper triangular is the number of wins for Viral A, and the value of lower triangular is the number of victory for Viral B

4.1 Simulation Results

We observed winning percentage of each company in 100 simulation trials under each pair of the strategy. The win-loss records are shown in Table 3:

1. If the content strategy is set as the trend strategy and targeting strategy is set as the degree strategy, the company can win the match in the high winning percentage no matter what strategy pair the opponent uses.
2. If the two companies choose the same content strategy, the rank of the winning percentage for attracting visitors for each targeting strategy is degree strategy > clustering coefficient strategy.
3. If the two companies choose the same targeting strategy, the rank of the winning percentage for attracting visitors for each content strategy is trend strategy > learning strategy.

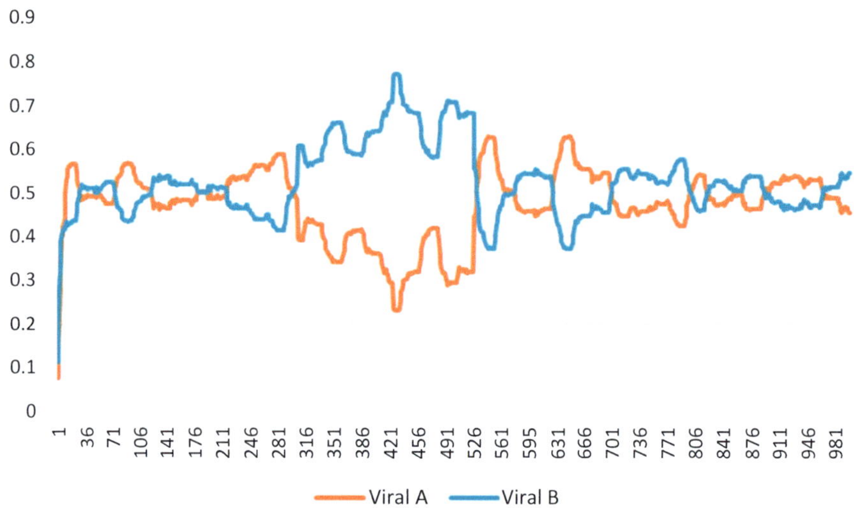

Fig. 4 The transition of the percentage of the latest customer visit

Table 4 The original company which produces the content

11th term	12th term	13th term	14th term	15th term	16th term	17th term	18th term	19th term	20th term	21st term	22nd term
A	B	B	B	B	B	B	A	A	A	A	B

From now on, we will focus on the competition between the two companies, Viral A and Viral B, using strategy pair (1) and (4) to analyze the detail of each strategy.

Figure 4 shows the transition of the percentage of the latest customer visit in the whole social population. From this graph, we can see that the share has rapidly changed from step 288 to step 528. We focus on this period and see what kind of contents each company delivers and how these contents attract the customers from SNS to their own website.

Table 4 shows the original company of the hit contents followed by media company B using the trend strategy as their content strategy from the 12th term (step 288) to the 22nd term (step 528). Figures 5 and 6 show the scatter plot of the similarity of delivered contents and user preference and the attracted customers through delivered contents.

From Table 4 and Figs. 5 and 6, we can observe that at the 12th term the media agent B imitated the contents delivered by media agent A at the 11th term. Furthermore, the media agent B delivers the contents which are highly similar with target users' preference from the 12th to 16th term and gain lots of customers. On the other hand, media agent A delivers the contents which have low similarity with target user preference from the 12th to 17th term and lost the market share rapidly.

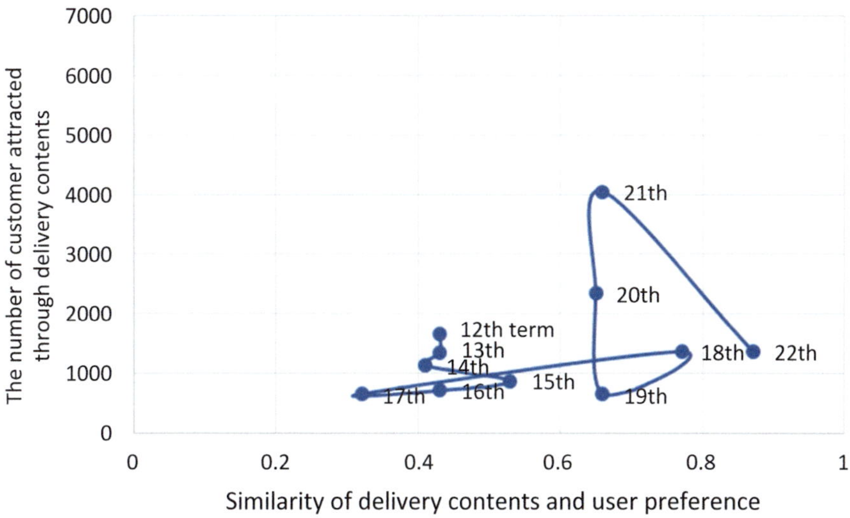

Fig. 5 The scatter plot of the similarity of delivered contents and user preference and the attracted customers through delivered contents (Viral A)

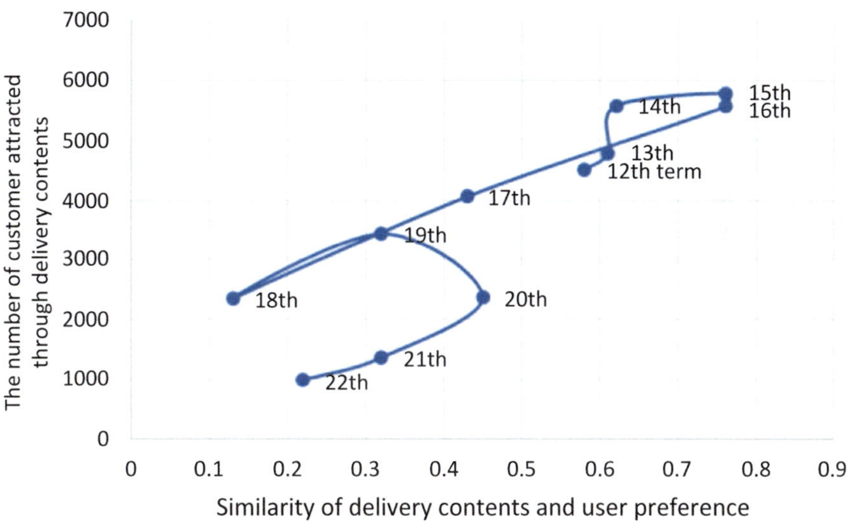

Fig. 6 The scatter plot of the similarity of delivered contents and user preference and the attracted customers through delivered contents (Viral B)

However, from the 18th to 21st term, the media agent A delivers the attractive contents and recovers the market share, but the growth has stopped at the 22nd term because of the imitation of media agent B using the trend strategy. From this result, we can say that if the company using trend strategy produces the attractive content itself, it will win the competition, and if the opponent company produces the attractive contents, it will imitate the opponent and block the increase of the opponent's market share.

So, we can conclude that if the market has the contents which can rapidly attract the customers, the trend strategy has a superiority in attracting more customers.

On the other hand, if the market has not delivered the contents which can attract lots of customers, the trend strategy does not have much superiority in attracting customers from SNS.

5 Discussion and Conclusion

In this paper, we present an agent-based model simulating the management process of viral media and attempt to find the best content selection and content targeting strategy for viral media to attract more customers. By comparing the two media sites' number of attracted customers in the competitive setting, we concluded that setting the content strategy as the trend strategy and targeting strategy as the degree strategy is the most effective strategy pair to gain customers from the social media. Meanwhile, we also successfully reproduce the phenomenon that the "plagiarizing viral media" has been increasing in real society. For the future work, we will increase the number of media agent to do the simulation. Also, the benefit and cost of the media agent should be taken into consideration in our model.

Acknowledgments Financial support from the Japan Society for the Promotion of Science (JSPS), Grant-in-Aid for Scientific Research (A) (25240048) (Terano) and Grant-in-Aid for Scientific Research (C) (15K01180) (Yamada), is gratefully acknowledged. We thank anonymous referees for helpful comments and suggestions.

References

1. BuzzFeedPress. (2013). BuzzFeed reaches more than 130 million unique visitors in November.
2. Bampo, M., Ewing, M. T., Mather, D. R., Stewart, D., & Wallace, M. (2008). The effects of the social structure of digital networks on viral marketing performance. *Information Systems Research, 19*(3), 273–290.
3. Rand, W., & Rust, R. (2010). Agent-based modeling in marketing (Working Paper).
4. Bäck, Thomas. (1996). Evolutionary algorithms in theory and practice. Oxford University Press, 120 p.
5. Hildebrand, C., Hofstetter, R., & Herrmann, A. (2012). Modeling viral marketing dynamics in social networks – Findings from computational experiments with agent-based simulation models. In: *International conference on information systems*, Orlando, FL.
6. Barabási, A.-L., & Albert, R. (1999). Emergence of scaling in random networks. *Science, 286*, 509–512.

Land Use Decision-Making Strategy in Bandung: An Agent-Based Modeling Approach

Ilham Fadhil Nurdayat and Manahan Siallagan

Abstract Land system science as a complex system has been explored by using various approaches such as remote sensing, economics, ecology, and geography. Agent-based modeling (ABM), as a third way of doing science, enables researchers to explore the complex system deeper on land science. Interdisciplinary approach has brought land use science into agent-based modeling. Bottom-up view and inclusion of agent as real-life representation brought a powerful method to analyze land use change and cover (LUCC) and explore coupling between human and natural system. Land use decision-making is a function of interaction between internal models of the land manager with its environment. This study incorporates two variables, commercial and farm expectation, as a representation between productivity farmland and other commercial objectives of the land. Agents have the ability to alter land use based on rational and irrational decision-making process. The objective of this paper is to capture the behavior of the agent and observe land use change that resulted from agent decision-making.

Keywords Agent-based modeling • Decision-making strategy • Land use change and cover

1 Introduction

Java Island, in the early 1990s, has been experiencing the beginning of change of land use, especially the agricultural land. City expansion caused to shift the agricultural land into other cover, such as settlement and industrial area. Economic growth has been a catalyst for urban growth in Java Island [1, 2]. The economic

---

I.F. Nurdayat (✉)
Institut Teknologi Bandung, Bandung, Indonesia
e-mail: ilham.fadhil@sbm-itb.ac.id

M. Siallagan
School of Business and Management, Institut Teknologi Bandung, Bandung, Indonesia
e-mail: manahan@sbm-itb.ac.id

U.S. Putro et al. (eds.), *Agent-Based Approaches in Economics and Social Complex Systems IX*, Agent-Based Social Systems 15, DOI 10.1007/978-981-10-3662-0_7

growth that centered in urban area demanded goods and services other than food. Food production moves farther relative to urban area. Average return of agriculture is lower than other opportunities in urban area; thus, farmers change their activities that resulted from land trade or land use change. Aggregation of those phenomena has driven lower agricultural production which in the long term becomes vulnerable to national food security [3]. In addition, stagnant productivity in agricultural production cannot increase regional and national production capacity. Study about land use change and cover (LUCC) is important to understand its land use impact and study the impacts.

Bandung is the capital of Jawa Barat province. Established in the colonial era, Bandung was developed to become a plantation hub. This plantation hub grew becoming the center of economic activity in present day. Based on the government's statistics agency, in the end of 2014, Bandung was inhabited by 2,470,802 people. This figure reflected a slight decrease from previous year, where 2,483,977 people were registered in Bandung area. Meanwhile, Bandung Regency recorded an increasing number of residents, from 3.415.700 in 2013 to 3.470.391 in 2014. A decreasing number of residents in Bandung city are accounted by the rapid development of city center, thus resulting in high rent land. Residential and mostly agricultural lands drive to move into fringe of Bandung city. This could have an effect on shrinking agricultural land, thus decreasing the food production [4]. Concentrated urban area, such as in Indonesia, needs to develop sustainable urban development. Sustainable food production is one of them. The study about the land conversion in urban area, especially in Bandung, would give insight to Bandung policy makers for a better understanding of the behavior of LUCC in Bandung area.

Previous approach has been focusing on macro level by analyzing the factor that affected the LUCC. This approach allows the researcher to study the detail of the factors and make decision based on that information. Hence, an intervention will be developed based on that factor. Land system science is needed for interdisciplinary approach to describe the situation [5, 6]. Interdisciplinary field concerning climate change, food security, urban planning, and ecological studies has been studied using various approaches such as mathematical modeling, econometrics model, and remote sensing. Most of the approach focuses on top-down approach of the subject. Each modeling approach has different purposes and goals [7]. Top-down approach advantage enables the decision-maker to make general action, although the dynamics of the situation is less captured in this approach. In complex situation such as LUCC, capturing the situation dynamics will improve the understanding of actual situation.

Agent-based modeling (ABM) on decision strategy can offer alternative explanation in LUCC. The dynamics of the situation of the agent-based model can give a richer picture to the land system science. The dynamics of the situation in LUCC is captured by the interaction among agents. This agent is defined as landowner whom can change the purpose of their land use based on their strategy in decision-making. Difference in decision-making strategy results from their mental model, availability of information, and expected utility value of each agent. This dynamic interaction aimed to be preliminary studies for household decision-making regarding their land use decision.

2 Literature Review

2.1 Complexity of Land Use Change and Cover (LUCC)

LUCC have different sources of complexity in relation to human land use decision from human decision-maker to potential global-scale influences [8]. Complexity science focuses on disorder, instability, and change. This complexity can emerge from nonlinear interaction among land's stakeholders. Complexity of the interaction aimed to capture the progression of LUCC based on landowner's decision. In human decision-maker level, individual land manager can employ different strategies. Individuals can make their own strategy based on their preferences, beliefs, information, behaviors, and expectations.

Recent development in land use science enables us to incorporate the behavior of human and the society, the multilevel characters of decision-maker and land units, and how are they connected to a broader world [9, 10, 11]. That opens the opportunity to study interdisciplinary aspect in land use. Previous studies have attempted to combine land use, geographic information system (GIS), human behavior [11], food security, and climate change [12]. This development opens the opportunities for social science to develop strong social simulation [13].

Previous model explicated the causality of the land use change by modeling the trade, supply, and demand that are connected by endogenous price mechanism. Deeper understanding of LUCC different approach. Land use change caused by supply and demand process is not included in the model. Remote sensing is used to focus on land supply and spatial pattern data. This model is beneficial when explaining about spatial limitation and land resources. But this modeling approach cannot capture endogenous supply, demand, and trade [10].

LUCC rely on the decision-making of the land manager for productive, rent, or other purposes. The decision is affected by the environment, economic opportunity, social aspects, and government regulation. This research focuses on farmers as land managers, because household farmer is the backbone of the agricultural production in Indonesia. Despite its important role, the numbers keep decreasing. From agriculture census data in 2013, the number of household farmer decreases 16, 32% compared to 2003. This event resulted in decreasing agricultural production.

2.2 Driving Forces and Actors

Previous studies focus on identification of driving forces in land use change. Ecological field has seen natural factors as driving forces especially that alter environmental condition [14, 15]. In the development, driving factors alone cannot cause land use change, in order to create decision of the land use. Driving factors need to interact with the decision-maker. Natural, politics, social economy, culture, technology, and society were the main driving factors in interdisciplinary studies of land use change. Each of the driving factors has different ways of influencing the decision-maker.

Interaction between actors and driving factors is summarized into four models [16]. The first (DF-C) model is that driving forces are directly related to land use change. Causal relationship is not of prime interest in this model. The second (DF-A-C) model represents that driving forces have an influence toward actors' motivation and therefore actors' actions. The third model (DFA-C) focuses on the close relation and interaction between actors and driving forces that caused land use change. Change occurs as a result of that interaction. The last model (AC) depicted that actors have central role in land use change; this model represents the situation where driving forces are elements of the environment in which actors make decision. Each type of model has a different approach to capture the situation. AC model type represents that actors have central role suitable with research objective. Agent-based modeling was appropriate to describe this type of model because in ABM the modeler focuses on the actor and actors' actions, motivations, decisions, and discouragements. Different types of actors result in heterogeneous and more dynamic interaction of the problem.

2.3 *Strategy of Decision-Making*

The strategy is a process of taking decisions by the agent, from how he gets his information to how to use that information to make decisions. The fundamental assumption of decision in microeconomics is that the people maximize their benefit and minimize the cost. This way of thinking is also described as rational decision-making. People assume they have adequate information and analyze it to get maximal outcome. The strategy is a process of rational strategies applied by taking into account the profit and loss as a result of the action.

On the other hand, other disciplines suggested people are rational decision-makers, even when they are facing a variety of limitations. Based on Kahneman's work, people cannot take rational decision if they are facing incidental risks, making them prone to intuitive decision. At the stage of decision-making, the rational assumption might be useful, but to see the dynamics resulting from interactions need enter the presence of irrational strategy.

In everyday interactions, the two strategies above may occur for personal decision-making depending on whether it has enough information and what mental model that he had. Lack of information can lead to people taking decisions based on the available data, even if it has no relevance. This paper aims to see how the dynamics between the two strategies in the related land use decision-making (Fig. 1).

3 Methodology

This research focuses on a high-level abstraction model by incorporating individual land manager strategies into the simulation model. This model was built using two decision strategies which constitute in this paper as "rational" and "irrational."

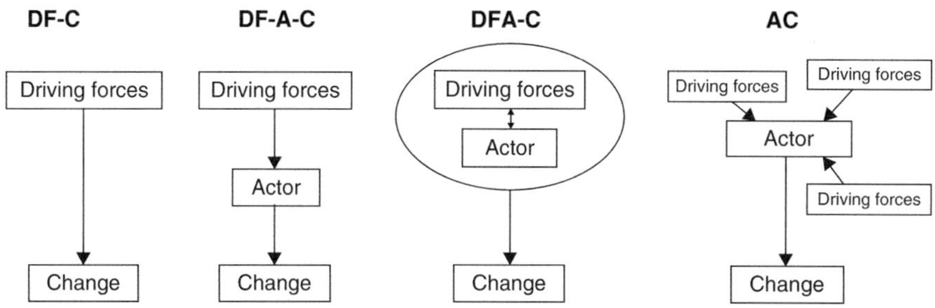

Fig. 1 Driving force and actor linkage

Rational decision-making strategy is described as decision-making strategy to maximize outcome of the landowner. Each agent has production function to produce to fulfill the demand. Irrational decision-making strategy is explicated as human behavior to make decision based on the rule of thumb and decide land use based on relative valuation, although it's not symmetric information. Several assumptions and limitations will be applied in this model; a more detailed explanation is included in simulation model section.

Agent-based modeling (ABM) is used in this study to capture heterogeneous condition in individual decision-making strategy. ABM has the advantage to incorporate the behavior aspect to agent decision-making strategy. These studies observe interaction between agents that have different decision strategies.

Individual decision-making is affected by extensive factors, internally and externally. Internal factors such as personal belief, culture, and past experience have been studied using the theory of planned behavior, factor analysis, and qualitative analysis. On the other hand, external factors, such as the environment, information access, and policy, have been studied in the macro-focused discipline. In microeconomics, the decision-maker is the center of studies. Individual behavior and action are theorized and hypothesized. The fundamental theory of microeconomics suggests that human decides based on rational aspects, in other words, maximize the benefit and minimize the cost. This hypothesis affected the decision-maker's view to develop an understanding of the problem and the solution of it.

4 Simulation Model

Overview, design concepts, and details (ODD) protocol was developed as structured design for an agent-based design process, to overcome barrier in the specification and explanation of the model. Thus, the model could avoid misunderstanding among other researchers and create bigger opportunity to replicate, refine, and redesign for further research [17]. Application of ODD protocol has not been widely used, but there are an increasing number of researchers attached to their ODD protocol in their paper. Critiques for the proposed ODD protocol were the methodology generated from ecological perspective, and some adaptation was needed to be accommodated that includes human decision-making process [18]. The organization of section will follow section and elements in ODD protocol.

4.1 Overview

4.1.1 Purpose

This model was intended to explore and understand land use change decision-making strategy. This research will look at agent's decision-making strategy regarding their land use. Decision-making strategy was divided into two types: "rational" and "irrational" ones. Both interact with each other in order to decide what decision they take for their land. Targeted users of this model are academicians, researchers, and decision-makers.

First, strategy is derived from farmer economic strategy in making decision to use their resource [3]. Land is assumed as one of the farmer's resources; thus, "rational" strategy was selected to represent landowner behavior. Second, strategy selection is based on basic human behavior to mimic other decisions with limited information. Further process will be explained in the next section.

Emergent properties of the model intended to see which strategies generate disparities among land use. Land use is simplified into two categories: the first is agriculture and the second is nonagriculture to support the model intention. Model intention does not make prediction or projection, but means to study decision strategies that have influence over land use decision of the land manager. Individual-level decision in aggregation will drive change in macro-level properties (Fig. 2).

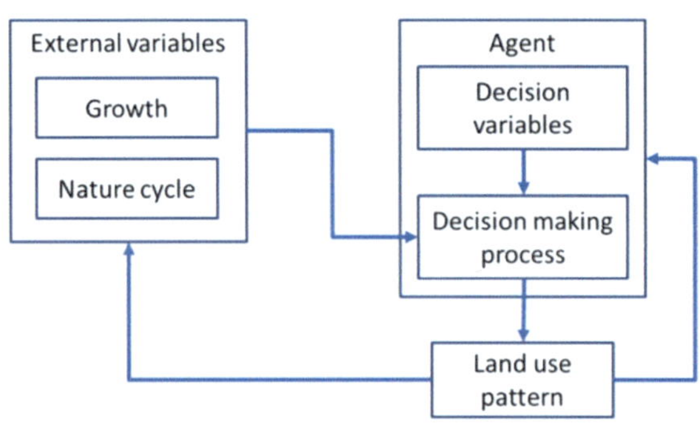

Fig. 2 Conceptual framework

4.1.2 Entities, State Variables, and Scales

The model will consist of two main entities: the environment and the agent. Current model represents human decision-making strategy. This model excludes spatial aspect and only focuses on decision strategy aspects. Agent will interact with the environment and other agents that consequently influence agent's decision-making attribute and process. Each agent is assumed to own a land without considering the dimension and characteristic of the land. Each agent has decision toward the use of their own land.

Landowner agents have state variables as follows (Table 1).

The environment has state variables as follows (Table 2).

4.1.3 Process Overview and Scheduling

Every time step, agent's land use decisions were determined by a decision-making strategy they were assigned. For agent irrational-type decision-making strategy, information sharing and asymmetric information were included in this process. Two spots were provided for the agent to exchange information within the spot. Information sharing is restricted to only in their spot. The spot in this model is named komunitas. Komunitas is an Indonesian language for community or in this paper defined as a place or activity for information sharing with other agents.

A probability of 0.5 is randomly assigned in the beginning of each time step to komunitas 1 or komunitas 2. Komunitas acts as an institution for agents to share their information about their productivity. Information sharing is represented by forming average productivity.

Table 1 Agent's state variables

Variable	Brief description
Productivity	Level of agent production rate given to resources
Capital	Starting and accumulated resource for production activity
Probability to change	Agent willingness to change, represented by a number between 0 and 1
Satisfaction	State of agent's satisfaction to their condition. Satisfied or dissatisfied
Decision strategy	Which decision strategy the agents chose. Rational or irrational

Table 2 Environment's state variables

Variable	Brief description
Demand	Demand that should be fulfilled by agent production
Average land use	Calculated average
Agricultural growth	Rate of agricultural land use resource
Nonagricultural growth	Rate of nonagricultural land use resource

$$\text{averageproductivity} = \frac{n_1 + n_2 + n_3 + \cdots + n_i}{\sum i\text{komunitas}} \tag{1}$$

$$\text{productionfunction} = \text{capital} \times e_i \tag{2}$$

$$\text{random}(x) < \text{probability} \tag{3}$$

Each agent will use average productivity as benchmarking standard. If their productivity is below average, they will increase probability to change the land use; otherwise, if their productivity is above average, they will increase the probability to unchange the land use.

Rational agent type has a different decision-making strategy; this type doesn't rely on other information to make decision. The agent relies on production activities that resulted from Eq. (2). The outcome of the equation will be used to fulfill external demand. If they cannot fulfill the demand, the probability to change land use will increase.

After the agent updates their probability, land use change process is triggered by random number generation between 0 and 1 to run Eq. (3). If probability of the agent is larger than random number, then the agent will change their land use from farm to commercial and vice versa. After each conversion the probability will reset to 0.1 to represent wait-and-see action after they convert their land.

4.2 Design Concepts

The emergence of this model is only represented by structural change in land use, excluding spatial aspects. Interaction between agent and driving forces holds central role in this research to capture an idea of how different decision-making strategies result in different land use patterns. Driving forces that shape land use are categorized into political, economic, cultural, technological, and natural. Economic and natural driving forces are included in the model. Economic forces in the urbanization argument are central to attract people and activities. Natural aspect represented resources of the land, and agricultural use accounted for the land quality. For each production activity, land quality decreases, but periodically land could update the resources, thus simulating season in Indonesia.

Agricultural type of land is used by the owner as a farm to produce crops or as a green space. This type of land has been decreased in effect of urban development that directly and indirectly pushes the land use change. Nonagricultural land includes land for commercial purposes such as service or manufacturing activities. This type of land has been increasing in urban area recently. But the generalization of nonagricultural land may be considered too broad, because residential lands were included in this type although this is not a productive type of land. Agricultural activity relies on the availability and the quality of the land as production factors, thus resulting in competition of land. Interaction among stakeholders affected governance structure, production, consumption, technology, ecosystem services, and global environmental change in human activity in regional and international scale.

4.3 Details

4.3.1 Initialization

Initialization phase begins with generating agent decision variables and external information. Decision variables for agents consist of economic and natural productivity. Selection of these variables was based on five categories of driving forces in land use change. Economic and natural driving forces were selected to represent competition between agricultural and nonagricultural land use. Further research is needed to comprehend this selection of decision variables.

This model begins with high level of abstraction. All agent variables are set to 0.2 to assume similar productivity in the land. Productivity will differ from several scenarios. Each productivity represents different types of land; economic productivity is for non-farmland which in the real world is used for commercial activities such as trade, manufacturing, and service. Natural productivity in this model represented agricultural production in farmland.

4.3.2 Scenario

Describing different decision-making strategies in this model was done using several scenarios in order to give clearer distinction to each process. The following are the scenarios that were implemented in this model (Table 3).

Scenario 1, 2, and 3 emphasis on irrational and rational decision-making strategy focuses on each decision-making strategy and how they interact with each other. The total number of land used for farm or commercial will be used as emergence properties to view these two decision-making strategies.

Table 3 Model scenario

Name	Description
Scenario 1	Irrational decision-making strategy
Scenario 2	Rational decision-making strategy
Scenario 3	Irrational and rational decision-making strategy

5 Result and Conclusion

In this section, we will describe and discuss the simulation result. First, we look at decision-making strategy change, and second, we analyze the relation between the two decision-making strategies in deciding land use.

In scenario 1, agents were assigned to have irrational decision-making strategy, and the result seemed the land use did not have distinctive patterns. They change their land use according to the information they have in the community, and the land manager is constantly adjusting the land use to the information they have.

In scenario 2, agents that are assigned to rational decision-making strategy inclined to convert their land into commercial land. This pattern emerges from the beginning part of the iteration. In Fig. 3 total land use number in scenario 1 showed that land use change rate is very rapid; in less than half of the total 200 iterations, almost all landowners have convert their land to commercial land (Figs. 4 and 5).

Different rules were adapted to natural and economic productivity to capture real-world condition. For natural productivity, each time step in this productivity will decrease for six periods for 0.1 per each time step, and then at the sixth time step, the productivity will rise to 0.3. This rule meant to capture farming cycle due to natural condition of soil quality, season, etc. On the other hand, economic productivity was designed to adjust to the demand with 5% standard deviation.

Scenario 3 incorporates between irrational and rational decision-making strategy. The result of this scenario captures the interaction between those. Total land use number in scenario 1 has emerged but with some volatility in all iterations. Although nonagricultural land type has been the most land use, the number of farmland in this scenario is significantly larger than scenario 2. This evidence could show us that

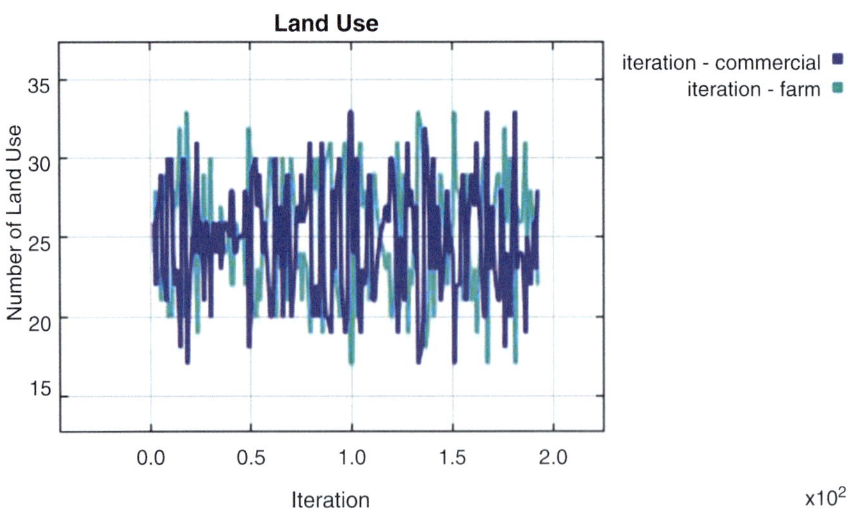

Fig. 3 Total land use number in scenario 1

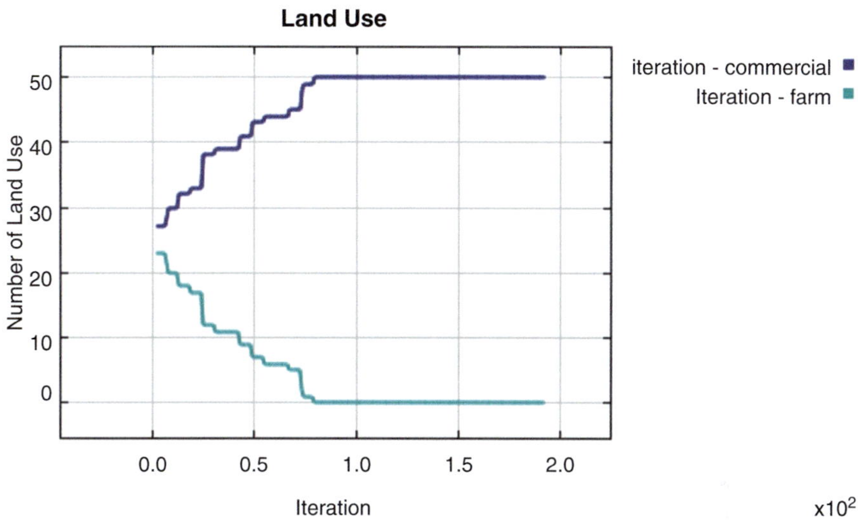

Fig. 4 Total land use number in scenario 2

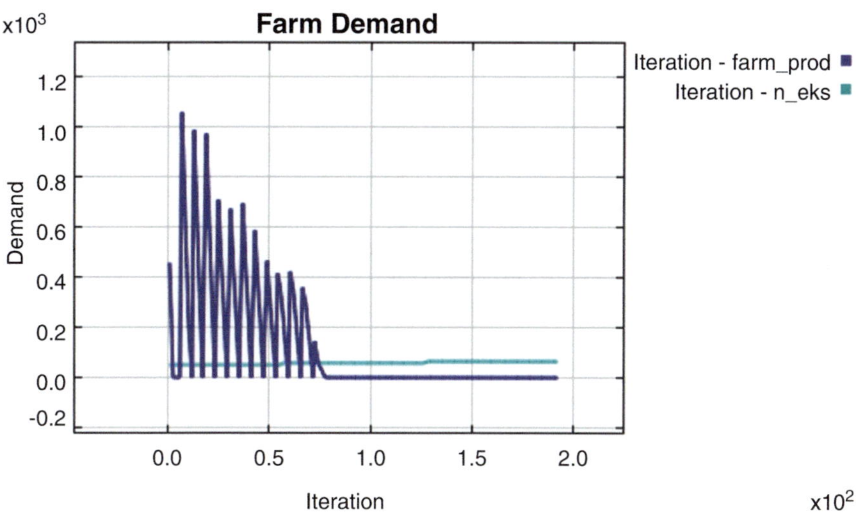

Fig. 5 Total land use number in scenario 3

the irrational decision-making strategy has increased farmland in the model. Main driving irrational and rational agent type sharing their information about their land use that increase volatility of the land use change and increase number of farmland (Fig. 6).

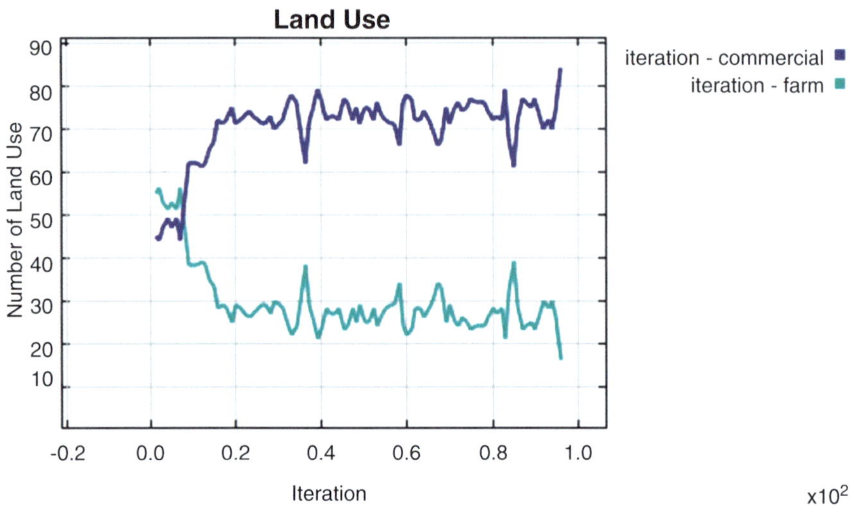

Fig. 6 Demand fulfillment comparison in scenario 2

This model showed in the broad sense that different decision-making strategies of the agent could result in different land use types and patterns. Irrational decision-making only relies on other information not regarding the truth and completeness of the information. This kind of agent decision-making strategy can be observed in the real world. In land market, rumor, gossip, and word of mouth have important roles to shape people's decision.

The opposite decision-making strategy was described in the rational strategy. The agent ignores information from other agents and relies only on external environment driving forces. Decision-making strategy has a close resemblance, based on our analysis, with the fundamental approach in capital market. Rather than using relative valuation tool, they studied about the business process and prospectus of the company. This decision-making strategy was considered more rational because the process agent could maximize the opportunity.

Economical and natural decision variables in this model are claimed as the beginning phases in the model building. Each driving force only affects one land type, and in the future several driving factors could be included in one equation to represent a decision-making result of intertwined several driving factors that cannot be separated. This view supports the fact that in Bandung, based on National Institute of Aeronautics and Space (LAPAN) satellite imaging, agricultural lands have been decreasing more than 90% since early 2000. Several studies argue the decreasing process is due to Bandung economic growth that shifts the main activities to factory, craft, and service. Increasing demand for settlement has driven up land price, thus shifting benchmarking process with agricultural production result. Short-term high land price creates incentive for land managers to change their land use for more optimum outcomes.

In understanding build complexity, we must build a complex model. But to build this model that could represent complex model is a long road we take step by step. Comprehension and a deep understanding of the model result in the better model. This is the main objective of this study: to develop high-level abstraction model to represent decision-making strategy, excluding spatial aspect in land use change. Decision-making strategy, the object of the study, is an important aspect to consider in land use change modeling. The pattern that emerges from land use change is not only caused by actor and driving forces but also from decision-making strategy in the individual level. Aggregation of individual level will create properties in, national and international stage. Agent-based modeling enables researcher to explore that particular objective and area of study.

Further research is needed to add more decision variables and driving forces into agents and decision-making strategy. Interplay between variables and different kind of decision making strategy in deciding land use change is important aspect in developing model and policy. Parameterization of data using real-world data to add verification process will strengthen the result of the study.

Bibliography

1. Firman, T. (2000). Rural to urban land conversion in Indonesia during boom and bust periods. *Land Use Policy, 17*(1), 13–20.
2. Verburg, P. H., Bouma, J., Veldkamp, A., & Bouma, J. (1999). Land-use change under conditions of high population pressure-The Java Case. *Global Environmental Change, 9*, 303–312.
3. Thamrin, F. D., Rizal, S., Bunasor, S., & Hari, W. (2015). Economic behaviour versus rice farmer environmental behaviour in making food independence: Case west Java-Indonesia. *Asian Journal of Scientific Research, 8*(1), 67–73.
4. Firman, T. (2009). The continuity and change in mega-urbanization in Indonesia: A survey of Jakarta–Bandung Region (JBR) development. *Habitat International, 33*(4), 327–339.
5. Turner II B. L., Lambin, E. F., & Reenberg, A. 2007. The emergence of land change science for global environmental change and sustainability, presented at the Proceedings of the National Academy of Sciences of The United States of America, vol. 104, pp. 20666–20671.
6. P. H. Verburg, K. H. Erb, O. Mertz, & G. Espindola 2013, October, Land system science: Between global challenges and local realities Current Opinion in Environmental Sustainability, *5*(5), 433–437.
7. Brown, D. G., Verburg, P. H., Pontius Jr., R. G., & Lange, M. D. (2013). Opportunities to improve impact, integration, and evaluation of land change models. *Current Opinion in Environmental Sustainability, 5*(5), 452–457.
8. Parker, D. C., Hessl, A., & Davis, S. C. (2008). Complexity, land-use modeling, and the human dimension: Fundamental challenges for mapping unknown outcome spaces. *Geoforum, 39*(2), 789–804.
9. Simon, H. A. (1955). A behavioral model of rational choice. *The Quarterly Journal of Economics, 69*(1), 99–118.
10. Rounsevell, M. D. A., Pedroli, B., Erb, K.-H., Gramberger, M., Busck, A. G., Haberl, H., Kristensen, S., Kuemmerle, T., Lavorel, S., Lindner, M., Lotze-Campen, H., Metzger, M. J., Murray-Rust, D., Popp, A., Pérez-Soba, M., Reenberg, A., Vadineanu, A., Verburg, P. H., & Wolfslehner, B. (2012). Challenges for land system science. *Land Use Policy, 29*(4), 899–910.

11. Jokar Arsanjani, J. (2012). *Dynamic land use/cover change modelling*. Berlin/Heidelberg: Springer Berlin Heidelberg.
12. Moore, N., Alagarswamy, G., Pijanowski, B., Thornton, P., Lofgren, B., Olson, J., Andresen, J., Yanda, P., & Qi, J. (2011). East African food security as influenced by future climate change and land use change at local to regional scales. *Climatic Change, 110*(3–4), 823–844.
13. Kohler, T. A., & Gummerman, G. J. (2001). *Dynamics of human and primate societies: Agent-based modeling of social and spatial processes*. Oxford University Press.
14. Aroengbinang, B. W., & Kaswanto. (2015). Driving force analysis of landuse and cover changes in Cimandiri and Cibuni Watersheds. *Procedia Environmental Sciences, 24*, 184–188.
15. Meyfroidt, P., Lambin, E. F., Erb, K.-H., & Hertel, T. W. (2013). Globalization of land use: Distant drivers of land change and geographic displacement of land use. *Current Opinion in Environmental Sustainability, 5*(5), 438–444.
16. Hersperger, A. M., Gennaio, M.-P., Verburg, P. H., & Bürgi, M. 2010. Linking land change with driving forces and actors: Four conceptual models. *Ecology and Society, 15*.
17. Grimm, V., Berger, U., Bastiansen, F., Eliassen, S., Ginot, V., Giske, J., Goss-Custard, J., Grand, T., Heinz, S. K., Huse, G., Huth, A., Jepsen, J. U., Jørgensen, C., Mooij, W. M., Müller, B., Pe'er, G., Piou, C., Railsback, S. F., Robbins, A. M., Robbins, M. M., Rossmanith, E., Rüger, N., Strand, E., Souissi, S., Stillman, R. A., Vabø, R., Visser, U., & DeAngelis, D. L. (2006, September). A standard protocol for describing individual-based and agent-based models. *Ecological Modelling, 198*(1–2), 115–126.
18. Müller, B., Bohn, F., Dreßler, G., Groeneveld, J., Klassert, C., Martin, R., Schlüter, M., Schulze, J., Weise, H., & Schwarz, N. (2013). Describing human decisions in agent-based models – ODD + D, an extension of the ODD protocol. *Environmental Modelling & Software, 48*, 37–48.

Agent-Based Simulations of Smallholder Decision-Making in Land Use Change/Cover (LUCC) Problem

Case Study: Agricultural Land Conversion in Jambi Province, Indonesia

Manahan Siallagan, Niken Prasasti Martono, and Utomo Sarjono Putro

Abstract Land conversion is one complex problem which includes actors and factors with different social levels. In the process of land use change, every small change in decision-making methods used by individuals may significantly affect the outcomes. Agent-based modeling and simulation (ABMS) is a common approach to analyze and simulate the process of land use change as the result of individual decisions. This paper firstly describes the general problem of agricultural land conversion (ALC) in Jambi Province, Indonesia, in particular in connection to the smallholders' role. It is then followed by the brief explanation on factors influencing the smallholders' behaviors. Then, conceptual framework of ABMS in analyzing land conversion as one kind of the land use change/cover (LUCC) processes is presented enriched with agent topologies and the decision-making processes on the agricultural land conversion problem. Finally the proposed model is illustrated through a preliminary case study in Jambi Province, and some scenario on the effect of interaction between farmers and government is described. The framework is still a general approach to analyze simple LUCC problem using ABMS approach.

Keywords Agent-based modeling and simulation • Decision-making • Food security • Land conversion • Land use/cover change

---

M. Siallagan (✉) • U.S. Putro
School of Business and Management, Institut Teknologi Bandung, Bandung, Indonesia
e-mail: manahan@sbm-itb.ac.id; utomo@sbm-itb.ac.id

N.P. Martono
School of Business and Management, Bandung Institute of Technology, Bandung, Indonesia
e-mail: niken.prasasti@sbm-itb.ac.id

U.S. Putro et al. (eds.), *Agent-Based Approaches in Economics and Social Complex Systems IX*, Agent-Based Social Systems 15, DOI 10.1007/978-981-10-3662-0_8

1 Introduction

In making decisions of the use of land, individuals face a complex environment that consists of interacting elements including both natural and human systems. The elements of natural and human systems that influence land use decision may be very complex. In the natural systems, land use decision-making mostly depends on the dynamic process of the nature, for instance, how the hydrological cycle works and how the distribution of biological resources applied in the specific area would differ from other area. Moreover, the social elements of human's environment can also influence the options and incentives that are available to the individual and the land use decisions they make.

Land use problems usually arise when there are significant changes in land use [1]. Land conversion often is associated with problems such as loss of farmland, food security problems, and other issues. There are a lot of land conversion types, but in this paper the focus will be on the conversion of agricultural land into land used for agro-industrial investments, especially on rubber and oil-palm plantation.

As the computer simulation technologies are developing, social scientists have provided a tool not only for predicting the social phenomena of land conversion but also for better understanding the nature of the human systems involved in the phenomena. For years, a number of agent-based simulations (ABMS) have been designed to capture the dynamics of land use and land cover change (LUCC). It has been a powerful approach in the land use modeling community mainly because it offers a way of replacing transition probabilities or differential equations at one level with decision rules at a lower level along with the appropriate environmental feedback [11]. These simulations have focused on both abstract and real-world examples in which the actions and interactions of a collection of agents, representing individuals, households, or organizations, determine the overall patterns of land use change produced by the simulations. An agricultural land use change model example –"FEARLUS" – investigates how well different social learning strategies are employed [12]. Further, "LUCITA" explores how the characteristics of frontier families influence changing agricultural land use, and secondary succession, in the rain forest [13]. Existing models and simulations typically couple a human system, represented by a collection of agents making land use decisions, with an environment system, represented by a raster grid of spatially distributed land uses within the landscape, through agent-agent and agent-landscape interactions that feedback and alter the LUCC in the area of interest.

ABMS approach is used in this paper to model the decision-making process of smallholders when facing the land conversion choice. In the model, land conversion behavior of smallholders would be mainly triggered by three factors: external factors, internal factors, and social factors. The rest of the paper is organized as follows: it firstly describes the general problem of agricultural land conversion (ALC) using an example of the case in Jambi Province, Indonesia, in particular in connection to the smallholders' role. It is then followed by the brief explanation on factors influencing the smallholders' behaviors. Then, conceptual framework of ABMS in analyzing land conversion as one kind of the land use change/cover (LUCC) processes is presented enriched with agent topologies and the decision-making processes on the agricultural land conversion problem. Finally the proposed model is illustrated through a preliminary case study in Jambi Province, and some scenario on the effect of interaction between farmers and government is described.

2 Land Conversion Problem in Jambi Province

Despite the fact that land conversion is a phenomenon that is almost unavoidable during economic development and population growth periods [4], uncontrolled land conversion will have great impacts on environment and agricultural products [5]. Not only in the developing countries had the case of agricultural land conversion (ALC), it also speeded in the developed countries.

Agricultural development basically has two principal objectives, to increase agriculture production both quantitatively and qualitatively and to increase the farmers' income. In the Ministry of Agriculture of Indonesia strategic plan (2010–2014), there are five objectives of the agricultural development [2]: first, to realize sustainable farming system based on the local resource; second, to establish sustainable self-sufficiency (*swasembada berkelanjutan*); third, to grow the food security and food diversification; fourth, to have more added value of the agricultural products; and fifth, the most important one is to increase farmers' revenue and welfare.

One of the problems faced by the nation is the ongoing conversion of agricultural lands to various other land uses, such as agro-industrial plantation that are economically more profitable. Other examples of the conversion on land use in other sectors are industrial area, public facility, and residential building. The agricultural land conversions are not only expanding in the agriculture production center such as Java Island but also in other region such as Jambi Province. As an illustration, there is a total of 75,000 ha agricultural land converted to oil-palm plantation in Jambi by 2010 [3]. The land conversion was triggered by the increasing market price of oil palm and the ease of getting the cash income periodically.

The decreasing number of agricultural land in Jambi Province was at least caused by two main factors: (1) the conversion in land use from agricultural to plantations, especially oil palm and rubber, and (2) the idled activity of land which is not cultivated for some period of time (idled land). Of the two factors, the main factor, namely, the transition function of land into oil-palm plantations and rubber, becomes the dominant factor. Many of the oil-palm and rubber plantations in Indonesia were established by large companies. However, smallholder farmers who are involved use more than 40% of the total oil-palm and rubber land [3]. Earlier, smallholder rubber and oil-palm cultivation were encouraged and supported through specific government support, but such policies are now terminated, and nowadays smallholders establish and manage their plantations independently. These land conversion decisions made by a large number of smallholders are more difficult to control than a large-scale land use transformation to companies.

3 Factors Influencing Smallholders' Behavior

In a paper examining behavioral change theory from an economic perspective, Prendergast et al. in the Scottish's government publication on agricultural and climate change [6] focus on three key drivers of behaviors: external factors, internal factors, and social factors. The factors would, if we apply it to the case of smallholder decision in the ALC phenomena, affect the decisions of smallholders, as social beings, including what to produce and how to produce it.

1. Internal Factors
   Based on [6], in smallholder farmer decision-making process, there are five keys to sum up in representing the internal factors: habit, personal capacity, framing and emotions, loss aversion, and immediate gratification and payoffs.
2. Social Factors
   Smallholder farmers' decisions are affected by the views and behaviors of their peers and neighbors as well as other family members and society at large [7]. Farmers are influenced by the behavior of their peer group. The literature shows that proficiently carrying out skilled farming improves both how farmers perceive themselves and how other farmers view them [7].
3. External Factors
   External factors are linked with monetary and effort costs – the affordability of choices, compared with the financial resources people have at their disposal, and the conditions which enable people to take advantage of these choices (such as accessibility or availability of information) or which act as barriers (complexity, inconvenience).

With reference to the aforementioned factors, we are interested in modeling the phenomena of agricultural land conversion done by smallholders with the following assumptions:

- *Smallholders are interested in the decision which offers an immediate gratification and payoffs.* In the previous works by Schwarze et al., it was stated that the oil-palm producers cultivate significantly more land than non-oil-palm farmers [8]. On average, oil-palm farmers cultivate 6.51 ha of land compared to 3.31 ha of non-oil-palm farmers, which is equivalent to almost twice the area.
- *Smallholders are affected by the behaviors of the neighborhood farmers.* The literature shows that proficiently carrying out skilled farming improves both how farmers perceive themselves and how other farmers view them [7]. Research by Diederen et al. [9] analyzes the choice of a farmer to be an innovator, an early (or late) adopter, and a non-adopter. The research found that structural characteristics explain much of the difference between types of farmer, and factors such as age and farm size and type may dictate whether and when adoption is a viable proposition at all. In addition, in the case of agricultural land conversion in Jambi, a survey found that smallholder farmers are aware to their neighborhoods' type of land, by means once the neighbor converts to the nonagricultural land, they have

the feeling of being insecure that their agricultural production will be possibly demolished by nonagricultural pests.

- *Smallholders who have sufficient economic support will tend to keep their agricultural land.* Smallholders argue that agricultural income has been insufficient in fulfilling the primary needs (housing, foods, and clothing) as well as the secondary needs (education, etc.) of their family [2]. To improve smallholders' sustainable livelihoods and agricultural practices, based on a baseline assessment and extensive stakeholder consultations carried out in 2011–2012 [10], the government of Indonesia is responsible for strengthening the smallholder aspect in such ways as strengthening local government agricultural extension worker systems to ensure sustainability and scaling up of successful solutions and working with smallholder cooperatives and larger plantations to reduce expansion into forests.

4 Purposed Mechanism

The model developed in this paper adapts the work of Robert Axelrod [10] known as the cultural dissemination phenomena. Axelrod models the adaptive model that reveals the effects of a mechanism of convergent social influence, based on the assumption that differences between individuals and groups exist. In his work, most neighboring sites have little in common with others and hence are unlikely to interact. However, when the two sites start to interact, they become similar and more likely to interact in the future. Over time, they share the same features, and it is shared over a larger area, and that represents how culture is disseminated over a group of agents [10]. The simulation model in this work is built using NetLogo (https://ccl.northwestern.edu/netlogo). Our purpose is to simulate the impact of smallholders' agent decision-making on the agricultural land conversion considering a simple behavior of the smallholders who own three factors (internal, external, and social) as an agent. The goal is to evaluate and compare the scenario's result, including some interaction with the government in supporting the smallholders in making the decision on ALC phenomena.

4.1 Agent and Environment

In the model, the only agents are represented by patches. Patches can't move, but otherwise they're just as "alive" as turtles [8]. There are two definitions of agents that will be used: smallholder agent and government agent. The variables owned by agents are listed below in Table 1, and the rule of interaction between agents is listed in Table 2, respectively. As agents, we defined that the farmers and governments are in a closed environment where their positions are randomly distributed. Though most of the works on modeling LUCC has engaged the spatial factor, in this model we focus on the social phenomena; thus, spatial considerations are excluded.

Table 1 Variables owned by patches

Variables	Definitions
Smallholder state	Smallholders' list of state of their own land
Government state	Government's list of features of their support
Money	Value of money to represent support and needs owned by smallholder and government agent
My neighbor	Neighbor of the patch

Table 2 Trait and value of each feature

Trait#	Trait	Value 1	Value 2
0	Type of land	Agricultural	Nonagricultural
1	Return	Less	More
2	Ease of process	Hard	Easy

Agent-based models consist of dynamically interacting rule-based agents. We model the smallholder agents based on these following rules:

When smallholder agents interact with other smallholder agents:

- One will be chosen to be active (random).
- Choose one of its smallholders' neighbors.
- Calculate choices similarity.
- With probability proportional to the choices' similarity, active site (smallholder) and the selected neighbor (smallholder) will interact with each other.
- Select a random feature on which the active site and its neighbor differ.
- Change the active site's trait on this feature with the neighbor's trait on this feature.

When smallholder agents interact with government:

- One will be chosen to be active (random).
- Choose one of its government neighbors.
- Calculate feature-choices similarity.
- With probability proportional to the feature-choices similarity, active site (smallholders) and the selected neighbor (government) will interact with each other.
- Check if there is any enough offer to be paid on the need of support. If it is, accept the support, by adding money with the value of the support and subtracting the same value to the government patch.

As for the government agent, agents are set to be interacting with smallholders only, and we model their rules as the following:

- Calculate feature-choices similarity.
- With probability proportional to the feature-choices similarity, active site (government) and the selected neighbor (smallholder) will interact with each other.
- Select a random feature on which the active site and its neighbor differ.

Table 3 List of each patch

	[2 2 1]	
[2 1 1]	[2 1 2]	[1 1 1]
	[1 1 2]	

Table 4 The percentage of traits of the middle patch

	33.33%	
66.66%	100%	33.33%
	66.66%	

- Check if there is any enough value to be paid on the need of support. If it is, give the support, by subtracting money owned with the value of the support and adding the same value to the smallholder patch.

4.2 Simulation Design

Each smallholder patch represents a state of each smallholder, and each government patch represents a feature of state of government support. The state of smallholder and state of government support of the agent are represented by a list of integers.

In this case, every patch owns a list containing three integers in the range of one to two. For example, here is a snapshot of four patches at a particular moment in Table 3.

Suppose the middle patch is one smallholder farmer, the inhabitant of the patch is a farmer who prefers nonagricultural land (2), less return (1), and easy process to get through agricultural products (2). Choices and feature similarity between two features can be defined as the percentage of traits they have in common. Table 4 represents the percentage of traits owned by the middle patch in common with each of its four neighbors.

5 Experiment Result and Discussion

5.1 Scenario 1: Smallholders Are Interacting with Smallholders, No Government Intervention

The simulation results are expected to illustrate the number of agricultural land and nonagricultural land (in unit) with different scenarios, while interaction only happens between smallholder farmers. Figure 1a shows the scenario results when there is no interaction between agents, so the number of agricultural land and nonagricultural land remains the same over a period of time. In Fig. 1b, we set the starting number of nonagricultural land more than agricultural land, and result shows that the number of agricultural land decreases over the period of time, while the nonagricultural land increases over the decreasing number of agricultural land. The list of feature is set to be random for the farmer agents. Lastly, Fig. 1c clearly

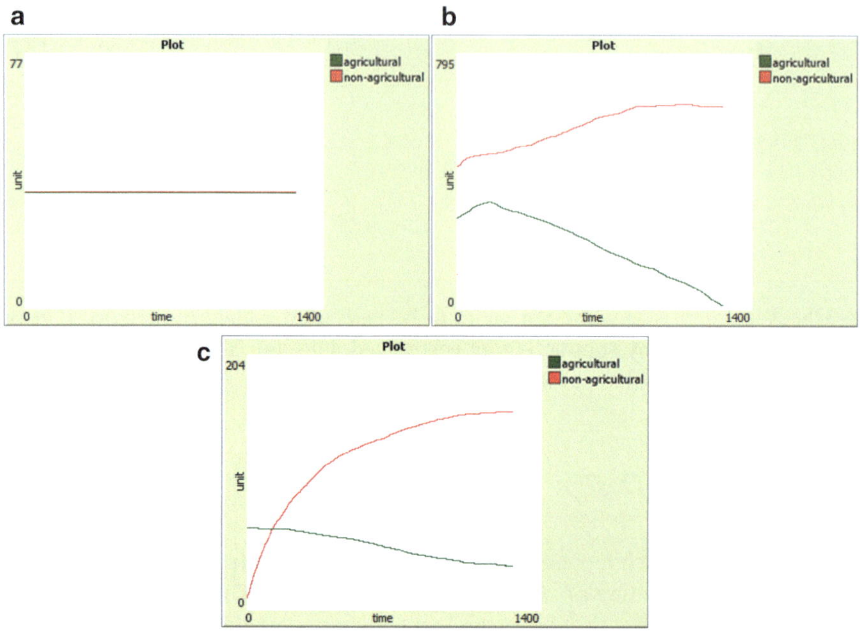

Fig. 1 Simulation results on smallholders' decision when (**a**) no interaction between other smallholders; (**b**) with interaction with other smallholders, with the number of nonagricultural farmers more than agricultural farmers; and (**c**) with interaction with other smallholders, with the number of nonagricultural farmers less than agricultural farmer

shows that the number of nonagricultural land will improve as agricultural land number decreases even when we set the starting number more. It summarizes that with the interaction between smallholder farmers and agricultural farmers tend to adapt to the nonagricultural farmer "culture," which own value "2" on the type of land trait with various values for other traits.

5.2 *Scenario 2: Farmers Are Interacting with Farmers and with Government*

The simulation results are expected to illustrate the number of agricultural land and nonagricultural land (in unit) with different scenarios, while interaction happens between smallholder farmers and also with the government. Government roles in this scenario are to offer their support (represented by number of money owned by each patch) and see whether the number of support needed by the smallholder can be covered by them to support them keeping their agricultural land. For the government agent, the traits are set to "1" value for the type of land traits. Figure 2a shows the scenario results when there is no interaction between agents, so the number

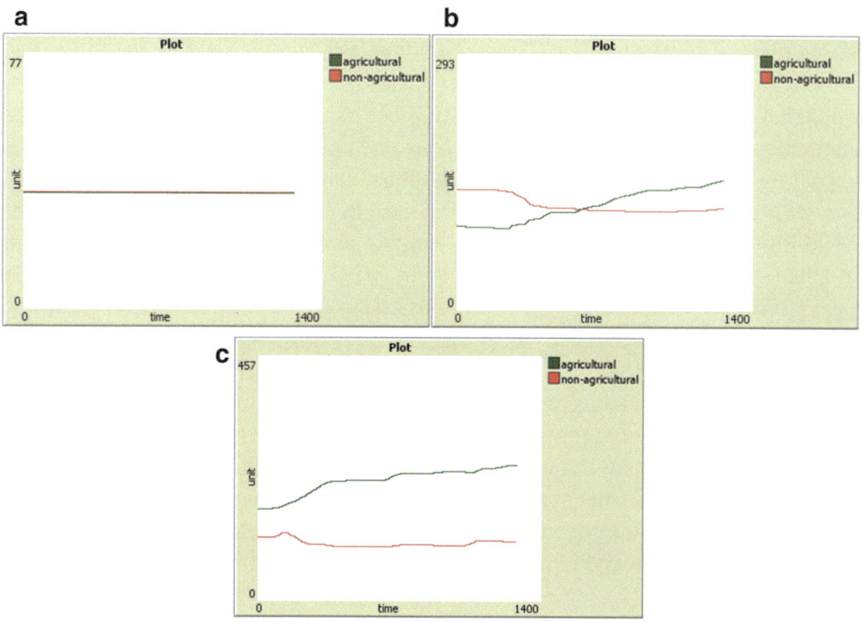

Fig. 2 Simulation results on smallholders' decision when (**a**) no interaction between other smallholders and smallholders with government agent; (**b**) with interaction with government, with the number of nonagricultural farmers more than agricultural farmers; and (**c**) with interaction with government, with the number of nonagricultural farmers less than agricultural farmers

of agricultural land and nonagricultural land remains the same over a period of time. In Fig. 2b, we set the starting number of nonagricultural land more than agricultural land, and result shows that the number of agricultural land increases over the period of time, while the nonagricultural land decreases over the increasing number of agricultural land. The list of feature is set to be random for the farmer agents. Lastly, Fig. 1c shows that the number of agricultural land will improve as the nonagricultural land number decreases by then. It summarizes that with engaging the government agent to the simulation, the smallholder tends to exchange trait with the government agent who also exchanges the number of value of money/support needed by smallholder.

6 Concluding Remarks

This paper use agent-based modeling and simulation (ABMS) approach to model a simple phenomenon on the agricultural land conversion (ALC) taking the Jambi Province as the case study. With several assumption on the smallholder agents about their driving factors of land conversion (internal, external, and social), we presented the result of simulation to see how interaction between smallholders and also between smallholders and government agent would have an impact on the number of converting land. Several findings are summarized by the following: (1) the interaction between smallholders without government (or other support) would have an impact on the increasing number of nonagricultural land, and (2) the interaction between smallholders with government (or other support) would have an impact on the increasing number of agricultural land. The simulation results go in line with the previous works on smallholders' behavior that in making decisions regarding land conversion, with some support and interaction with government, smallholder farmers might want to keep their agricultural land as long as they would have profitable production or they have basic support to avoid their land from idle production.

The work on this paper is a simple preliminary model on smallholder decision-making on land use change/cover process, but in the future, it is expected to be expanded with engaging more factors in the modeling process such as more agents to be included in the environment and more variables including geographical and spatial data to improve the validity of result to be implemented in the real world. Moreover, scenarios on government support would be improved by adding more value and possibility owned by the agent (not limited to economical support).

References

1. Journal, L. T. (2010). The driving forces of land conversion towards a financial framework for better land use policy. *Land Tenure Journal, 1*, 61–90.
2. N I. Minsyah; Endrizal E.E. Ananto F. Kasryno. (2014). Perubahan Luas dan Alih Fungsi Lahan Pertanian Serta Permasalahannya di Provinsi Jambi. In *Konversi Dan Fragmentasi Lahan Ancaman Terhadap Kemandirian Pangan*.
3. Göttingen, G., Euler, M., Schwarze, S., Siregar, H., & Qaim, M. (2015). Oil palm expansion among smallholder farmers in Sumatra, Indonesia, 8.
4. Tan, R., Beckmann, V., Van den Berg, L., & Qu, F. (2009). Governing farmland conversion: Comparing China with the Netherlands and Germany. *Land Use Policy, 26*, 961–974.
5. Azadi, H., Ho, P., & Hasfiati, L. (2011). Agricultural land conversion drivers: A comparison between less developed, developing and developed countries. *Land Degradation and Development, 22*(6), 596–604. doi:10.1002/ldr.1037.
6. Agriculture and climate change: Evidence on influencing farmer behaviours. (n.d.). Retrieved from http://www.gov.scot/Publications/2012/10/9121/6
7. Burton, R., & Dwyer, J., et al. (2006). *Influencing positive environmental behavior among farmers and land managers – A literature review*. Countryside and Community Research Institute and Macaulay Land Use Research Institute, Aberdeen, Scotland.
8. Schwarze, S. & Euler, M., et al. (2015). Rubber vs. oil palm: An analysis of factors influencing smallholders' crop choice in Jambi, Indonesia.
9. Diederen, P., van Meijl, H., Wolters, A., & Bijak, K. (2003). Innovation adoption in agriculture : Innovators, early adopters and laggards. *Cahiers D'économie et Sociologie Rurales, 67*, 30–50.
10. Axelrod, R. (1997). The dissemination of culture – A model with local convergence and global polarization. *Journal of Conflict Resolution, 41*(2), 203–226.
11. Verburg, P. H. (2006). Simulating feedbacks in land use and land cover change models. *Landscape Ecology, 21*, 1171–1183.

12. Izquierdo, L. R., Gotts, N. M., & Polhill, J. G. (2003). FEARLUS-W: An agent-based model of river basin land use and watermanagement. In M. Dijst, P. Schot, & K. de Jong (Eds.), *Framing land use dynamics: Integrating knowledge on spatial dynamics in socio-economic and environmental systems for spatial planning in western urbanized countries* (pp. 163–165). Utrecht: Faculty of Geographical Sciences, Utrecht University.
13. Deadman, P., Robinson, D., Moran, E., & Brondizio, E. (2004). Colonist household decisionmaking and land-use change in the Amazon Rainforest: An agent-based simulation. *Environment and Planning. B, Planning & Design, 31*(5), 693–709.

Understanding the Learning Processes of Traveller Behavioural Choices Using Agent-Based Approach: A Conceptual Framework

Yos Sunitiyoso

Abstract This paper presents a conceptual framework of agent-based model to study the learning processes of traveller behavioural choices. Social interaction and social learning among travellers are taken into the model. The way an agent-based model can be combined with traffic microsimulation model is also presented.

Keywords Agent-based model • Social interaction • Social learning • Travel behaviour

1 Introduction

Traffic congestion persistently becomes a major problem in transport. The widely quoted figure, which has also been criticised for being overly estimated (e.g. [13]), of annual cost of £20 billion and would increase to £30 billion by 2010, shows the importance in tackling this chronic transport problem. There are three different sources of congestion, namely, (1) recurrent congestion due to sheer weight of traffic, (2) congestion due to road works and (3) congestion due to incidents [11]. The first source of congestion contributes the most with 66% proportion (in 1998/1999), followed by traffic incidents 24% and then road works 10%. However, these sources of congestion are interrelated. A traffic congestion incident may be due to a combination of these sources. Especially when traffic flows approach too close to capacity, any of transient incidents and problem (e.g. roadworks or accident) will have a disproportionate effect [13]. At very low levels of traffic volume ('free flow'), changes in the number of vehicles have little effect. But as traffic volume increases, even very small increases or reductions in traffic will have a disproportionately large effect on speed.

Since travel time is still thought as the main consideration in travel decision-making instead of travel distance (e.g. [1]), the occurrences of traffic congestion

Y. Sunitiyoso (✉)
School of Business and Management, Bandung Institute of Technology, Bandung, Indonesia
e-mail: yos.sunitiyoso@sbm-itb.ac.id

U.S. Putro et al. (eds.), *Agent-Based Approaches in Economics and Social Complex Systems IX*, Agent-Based Social Systems 15, DOI 10.1007/978-981-10-3662-0_9

would definitely have a major impact on travellers' route decisions. When changes of traffic conditions occur causing travel delays, travellers will learn from these experiences. Studies found that travellers are more likely to switch routes when they know in advance of the trip that a route will cause them to arrive late (e.g. [26]). However, they often have route habits that produce inertia to stay with familiar routes [28], which may become barriers for travellers to change their route choices.

One of the critical issues in being able to predict traffic congestion levels is a clear understanding of the interrelated travellers' choices of travel mode, departure time and route through the transport network and how these choices can be influenced to achieve the greatest levels of both mobility and sustainability. This detailed understanding is necessary because the decisions made by each traveller in the system impact directly or indirectly on other travellers through road capacity and established speed-flow relationships. The understanding becomes more important when travellers are faced with permanent changes (e.g. new traffic signals, new bus exclusive lanes, congestion charging) or short-/long-term temporary changes (e.g. road or lane closures caused by minor or major works) in interurban or urban traffic networks, which may change their travel choices. Before changing their travel choices and making adaptation in response to the intervention or interruption in traffic network, there will be a time lag between when they know about it and when they actually change their travel choices. The time taken by travellers to adapt to (or to learn) the changes in traffic network will create inefficiencies in the system as traffic congestion will be generated as travellers have not (or could not) learnt about other travel choice alternatives. Thus, it is essential that attempts must be done to 'speed up' or to shorten the learning process or learning time.

2 Existing Research Context

2.1 Understanding Learning Processes of Travellers' Choices

The fact that travellers' choices change over time also means that there are learning processes of travellers when they are faced with changes in environment. Traditional assumption ignores that travel behaviour is a result of complex and dynamic processes involving interactions of many aspects and sequential adaptations over time [10]. It also ignores that current travel behaviour results from past behaviour since past choices have also influences on the choices at present [12].

Since the decisions made by each traveller in a transport system impact directly or indirectly on other travellers through their effect on capacity and speed-flow relationships, understanding travellers' interrelated choices of departure time, route and mode of travel (private car, public transport, cycling, walking, etc.) becomes critical to our ability to understand and predict patterns of traffic congestion. The evolution of knowledge is a key issue to be addressed, with three key mechanisms to be investigated: personal experience, social information/networking and official information.

By repeatedly making decisions, an individual acquires knowledge (learns) from *personal experience* about the environment and forms expectation about attributes of the environment. Individuals cognitively update their expectations and then store and retrieve information in their memory for the next decision-making. Transport system is highly dynamic, nonstationary and uncertain. Moreover, travellers' information is limited, imperfect and sometimes biased. Individuals do not know which choice is the best of his interest, so they decide to explore different choices in the beginning and become involved in more goal-oriented behaviour at later stage. However, this way is only effective if the environment is stationary. As most transport environments are nonstationary, travellers may try different choices occasionally and later to develop an adaptive behaviour with the process of learning. The concept of *individual learning* suggests that individuals learn from their experience and utilise an adaptive decision-making process to cope with uncertain situation (for a review, see [3]).

The effects of *social information* on travellers' behaviour are relevant to the existence of social interaction and social learning between them. These have been used in several contexts of transport studies particularly in travellers' responses to transport interventions in reducing car use. Using simulation studies, Sunitiyoso et al. [29] found that social interaction, which enables travellers to communicate with other travellers regarding their choices in different domains of social network (e.g. neighbourhood, workplaces, schools), increases the level of participation in car sharing. In the context of telecommuting decision, Paez and Scott [25] using two waves of simulation suggested that some marginal adopters of telecommuting are influenced heavily in the second wave by the decisions of others in the first wave. Kurumatani [18] introduced another type of information that can be provided to road users which is called a mass user support, which is a mechanism to coordinate users' intentions and preferences socially using cooperating software agents in a real environment. For example, in their proposed route information sharing (RIS) system [33], each vehicle transmits route information (current position, destination and planned route to the destination) to a route information server, which estimates future traffic congestion using this information and feeds its estimate back to each vehicle. Each vehicle then uses the estimation to replan their route. The study revealed that average travel time of drivers using the RIS system is substantially shorter than that of drivers who chose shortest distance or simple shortest time estimates.

Studies of the effects of providing *'official' or centralised information* (e.g. via advanced traveller information system – ATIS) found that providing travellers with such information do not necessarily produce benefits. Three classic terms coined by Ben-Akiva et al. [6] may exist: oversaturation, overreaction and concentration. If people are confronted with too much information, they are *oversaturated*, tend to ignore the information and develop simple heuristics to solve the problem. Another problem is the *overreaction*, which occurs if too many drivers respond to the information. Suddenly, congestion may be transferred from the original area to the alternative routes. The *concentration* effect happens when a great number of drivers select the best alternative route advocated by ITS, and consequently drivers with similar preferences will choose the same route, leading to congestion. However, as more intelligent devices will provide even more information about link travel times, densities or route guidance to the road user, these effects may be reduced with the availability of devices which provide more route alternatives rather than the best available route. The understanding of travellers' responses and learning to such information will be crucial for the development and effectiveness of such systems (for an overview, see [6, 22]).

Recent studies have attempted to study the effects of providing travellers with route information. Bazzan and Klugl [5] studied the effects of providing route recommendation to drivers in route choice situation where a Braess paradox exists. (Note: The Braess paradox is a phenomenon where adding a new road to a traffic network may not reduce the total travel time, but in fact, some road users may be better off but they contribute to an increase in travel time for other users.) They revealed that it is more beneficial for everyone to manipulate the route information given to the road users rather than providing them with the most accurate information (actual state of the system). Dia and Panwai [8] studied commuters' route choice behaviour in response to several types of official information (prescriptive, predictive, quantitative and qualitative). The simulation results support the notions that commuters' decisions to divert to alternate routes are influenced factors including the degree of familiarity with network conditions and the expectation of an improvement in travel time that exceeds a certain delay threshold associated with each road user.

2.2 Modelling Travellers' Choices Dynamics and Learning

In reality travellers' behavioural decisions are made based on the extensive but finite knowledge possessed by each individual traveller. However, the two common approaches to modelling such systems do not adequately capture the dynamic and reflexive status of such knowledge. *Macroscopic approaches* consider travellers to have 'perfect knowledge' of all travel conditions (flows, speeds, densities and how these change through time) and seek to achieve an equilibrium 'assignment' of the travel demand over the network. Some *microscopic approaches* consider travellers to have 'imperfect knowledge' of travel conditions, to allow responses to unexpected events (e.g. road accidents) to be represented. However, neither approach allows for travellers' knowledge to evolve over time as a consequence of their travelling behaviour.

To illustrate the consequences of this limitation on the prediction of congestion, consider the simple situation where a road is closed to create a pedestrian precinct. The macroscopic approach might produce the expected long-term assignment pattern characteristic of the system once it has settled down from the travel disruption caused by the closure, but it could not represent how the travel patterns evolve from the original pattern to this new situation. The microscopic approach might be able to consider how travellers would respond immediately to the closure but could not represent how this pattern would then evolve as travellers achieved a clearer understanding of the new traffic conditions. The example is simplistic in that it assumes that traffic conditions eventually reach a stable overall pattern. In reality continual small changes to both the underlying network infrastructure and the desires of travellers using the network mean that road systems are always in a state of evolution.

Most microsimulation models have focused on problems of short-term forecasting (e.g. estimating traffic flow impacts of some measures). They simulate traffic conditions based on equilibrium assumptions and often do not take into account the travellers' behavioural changes and learning process (e.g. drivers changing their travel decisions on subsequent trips in response to their new experiences of traffic conditions). As an exception, Dynamic Route Assignment Combining User Learning and Microsimulation (DRACULA, [19]) has attempted to incorporate travellers' learning into traffic microsimulation software by modelling drivers' day-to-day dynamic. Its learning model updates the experience of individual drivers and stores the information in their travel history files which influence their next day's choices. However, the learning model is a centralised mechanism performed by a single module over all drivers [27], and it is still quite simple (e.g. average of the last remembered experience). Any reference to drivers' histories or choices made during the simulation relates to the fixed pool of potential travellers who keep their identification through the simulation rather than each individual traveller.

Modelling approaches in complexity sciences enable the components of a system to be represented as distinct entities (known as agents) and importantly for this research allow those entities to change over time by adapting to their surroundings or as a result of specific information being supplied to them. In this manner the agent-based modelling approach can closely mimic the real-life situation, where individuals make what they feel are the best choices based on learnt knowledge that they have, allowing better prediction of the overall information (in-vehicle 'Sat Nav' systems, real-time passenger information, etc.) and considering the behavioural choices of travellers faced with future time-variant road charging experience.

In general, traffic systems, like urban traffic or pedestrian crowds, consist of many autonomous, intelligent entities, which are distributed over a large area and interact with each other to achieve certain goals. They are all changing actively the situation the traffic system is in. The concept of an agent is well suited for the description of road users in traffic scenarios. They are autonomous entities which are situated in an environment, adapt their behaviour to the dynamics they perceive (reactive) and interact with other agents (social) in order to achieve a specific goal (rational). The road user permanently perceives information via sensors and then reasons about it, makes a decision and acts on the environment via effectors [32].

The context in which the agent concept has been developed is usually considered to be a collection of 'self-contained problem solving systems capable of autonomous reactive, pro-active, social behaviour' [21]. Such 'multi-agent systems' are not models but problem-solving systems (O'Sullivan and Haklay [24]). In this context, agents incorporate sophisticated artificial intelligence (AI) techniques whereby they learn new ways to attain their goals. As an example, (software) agents are employed in information retrieval for some human client. Thus, they might take on the task of trawling the Internet for information likely to be of interest to its human client. In pursuing this goal, the agent may learn its client's preferences and also encounter other agents and communicate with them. We called agents in this context as *software agents*. Thus, not surprisingly, most studies were focused on applying the agent-based techniques to control systems and traffic management to make those systems more autonomous and responsive to recurrent traffic demand, for example, in traffic management (e.g. [17, 20, 31]) and route guidance system (e.g. [2, 16]).

Another context of the agent concept is in modelling human behaviours. Typically, an agent-based model has a large number of agents interacting in an artificial world representing some real-world location or social institution (e.g. a market). The modeller programme agents with plausible rules governing their behaviour and examines model outcomes to obtain insight into the real-world situation modelled. This agent-based model involves three basic ingredients: agents (which are 'people' of artificial societies), environment or space (which is a medium separate from the agents, on which the agents operate and with which they interact) and rules (which govern the way in which agents behave, as well as the way in which the environment may change over time). Agents in this context are called *behavioural agents*. This context has been used to obtain a better understanding of dynamic processes of travellers' decision-making and to make predictions about the occurrence of certain events (e.g. traffic congestions). And it also helps in developing theories by formalising theories and testing via simulation. Increasing effort has been dedicated to representing drivers' route choice and departure time behaviour and its underlying decision-making mechanism (e.g. [9, 27]). These works focus on practical implementation of a cognitive or learning model (e.g. BDI, beliefs, desires, intentions) to based decision-making carried out by driver/traveller agents. The behavioural agent context has also been applied in mode choice behaviour (e.g. [29]) and studies about the impact of ITS systems (e.g. [32]).

3 Conceptual Model

Travellers' choice making and behaviour can be considered as dynamic processes, since individuals can and do change their behaviours over time. The understanding of how travellers learn, develop and change behaviour over time is important in order to predict traffic congestion. A traveller's decision may be due to new information gained from their own experience, information and influence from the experience and behaviour of others through social networking and official information from the authorities. These three types of information may be able to accelerate travellers'

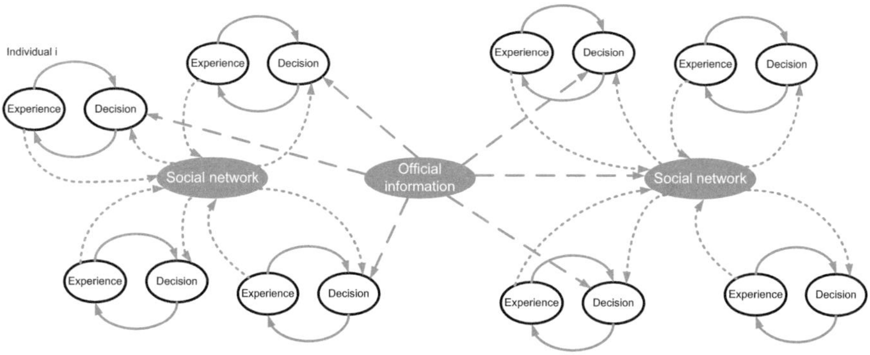

Fig. 1 Decision influenced by personal experience, official information and social network

learning processes or shorten learning time required to make adaptation. They are also interrelated to each other and may be combined to make decisions. For example, official information from authorities may be spread to other travellers within a traveller' social network and combined with the traveller's own experience to make a travel choice. In this research projects, we aim to understand whether and how these types of information influence travellers' learning and decision-making.

Personal experience about the consequences of their decisions (e.g. the travel time that results from a particular choice of route) is essential within each traveller's learning process. The way these experiences might be used to inform travellers' future behaviour involves an explore/exploit trade-off, where the tendency of travellers to have favoured behaviours will need to be represented. *Social networking*, where travellers discuss their personal experiences of traffic conditions with each other, are an important aspect of traveller knowledge that has been little studied. Social learning is distinct from personal experience in that it allows travellers to build a wider (although possibly biased) view of the network without having to experience the conditions themselves and therefore enables a faster accumulation of knowledge and a consequent acceleration in travel pattern evolution. *Official information*, often in the format of online congestion maps, can provide accurate information on travel conditions across the whole network but normally only at a high level of data aggregation and only for key routes. How travellers use this 'actual' knowledge in combination with their personal and social knowledge has yet to be understood (Fig. 1).

Representing the evolution of travellers' behavioural choices has received special attention from technical and scientific communities. Existing microscopic approaches suffer from this shortcoming, and the need for more robust and expressive modelling such as an agent-based approach is recognised. Agents will represent individual travellers each with their own explicit knowledge of transport network conditions and able to make their own travel decisions mediated by this knowledge. In this manner the agent-based modelling approach can be used to explore the theoretically motivated aspects of real transport systems that

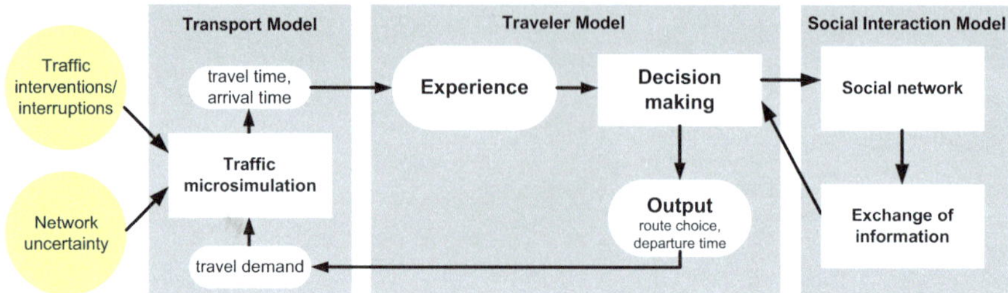

Fig. 2 Agent-based model

may otherwise be neglected. Importantly, it will typically be necessary to build idealised mathematical models alongside this agent-based simulation in order to gain principled understanding of its behaviour.

Individual learning models, which are based on theories such as expected utility models (e.g. heuristics, average return, weighted average return), prospect theory and fuzzy theory, and social learning models (e.g. imitation, confirmation, conformity models – [23]) will be incorporated in the agent-based model to simulate travellers' learning process. The types of official information could be actual traffic situation, route recommendation, manipulated information (e.g. [8]) or coordinated information for social coordination (e.g. [18]). Interaction between travellers will not only be modelled within the traffic microsimulation through speed-flow relationships but also be modelled through communication and exchange of information within social networks between travellers. Therefore, travellers will be able to learn not only from their own experience and/or from official information given by the authorities but also from other travellers' experiences (Fig. 2).

Basically, there are three main parts of the simulation model: traveller model, social interaction model and traffic model. *The traveller model* facilitates the decision-making processes of each individual traveller in making a travel choice decision (e.g. route choice, departure choice) by using both individual and social learning mechanism. Individual learning uses the traveller's personal experience, while social learning uses information gained from the social interaction model. Both empirical and normative models of individual and social learning, such as reinforcement learning, payoff-based learning, confirmation and conformity, will be tested in the simulation. The *social interaction model* facilitates the interaction among travellers through their social network through either direct interaction (e.g. face to face, phone, message) or indirect information (e.g. facilitated through social media). *The transport model* simulates space-time trajectories of individual travellers according to driver behaviour models through a traffic microsimulation, where its results such as travel time will become an input to add to the traveller's experience. Traffic interventions/interruptions and network uncertainty are controlled parameters that will be introduced in the traffic microsimulation according to the scenarios set for running the simulation.

This research project will not attempt to develop a new simulation package which will be able to do all these modelling purposes, instead it will also make use and develop upon existing models available to complete the objectives of the research, particularly the use of widely used traffic microsimulation softwares (e.g. PARAMICS, VISSIM) as the traffic model. This means that we will combine the agent-based traveller and social interaction models with these commercial softwares. However, recent development of a 'full' multi-agent-based traffic microsimulation tools (e.g. MATSIM, TRANSIMS) provides us with opportunity to develop our models utilising these modelling toolboxes. There are some potential advantages and disadvantages of these two approaches which will be discussed in the next two subsections.

3.1 Using 'Full' Agent-Based Traffic Microsimulation Approach

There are currently two toolboxes for modelling (large-scale) agent-based transport microsimulations which are available as open sources, TRANSIMS (http://transims-opensource.org/) and MATSIM (http://www.matsim.org/). Both toolboxes utilise agent-based microsimulation techniques and activity-based approach. They are quite similar except some differences such as the file formats, the handling of agents and activity chains and the speed of traffic flow simulation and traffic assignment routines [4]. These two models attempted to provide a new approach to travel demand forecasting to replace the conventional four-step models.

For further discussion we will focus on one of the toolboxes, MATSIM. The main advantage of utilising this toolbox in relation to this research project is in its dynamic traffic assignment which is made completely agent based. In particular, it is capable of feeding back agent-based (i.e. individualised) information, not only link-based information. The traffic flow simulation assumes the role of a realistic representation of the physical system, including explicit modelling of persons walking to the bus stop or of a bus being stuck in traffic. It is also possible to make each traveller react individually to exactly the conditions that this traveller has experienced, rather than to aggregate conditions. Some of its advantages includes its modular approach, so that MATSIM's modules can be combined, used stand-alone or replaced by customised implementations to test single aspects of a research work, and it supports large-scale simulation (e.g. several millions agents or network with hundreds of thousands of streets). Some disadvantages, compared to commercially available traffic microsimulation, may include its level of details (e.g. in models of driving behaviour, visualisation may not be as appealing as commercial microsimulation software (although impressive graphical visualisation of traffic does not necessarily imply an accurate simulation)), less user-friendly and more appropriate for large-scale simulation.

3.2 *Combined Multi-agent Learning Model with Commercially Available Traffic Microsimulation Model*

In recent years, increases in computing power have enabled more practical use to be made of microsimulation traffic models. These can be used to assess the effects of various measures applied to the network, such as ramp metering, route diversion, variable speed limits, travellers' information systems and route guidance. An essential property of all microsimulation traffic models is the prediction of the operation of individual vehicles in real time, over a series of short time intervals, and using models of driver behaviour such as car following, gap acceptance, lane changing and signal behaviour theories, rather than aggregate relationships. Although some microsimulation softwares (e.g. VISSIM, Paramics, AIMSUN) have been built based on representation of each vehicle-driver unit as an individual entity, they still do not allow drivers' knowledge or behaviour to evolve over time.

Among various traffic microsimulation softwares available, the most widely used (commercial) models in the UK [7] are VISSIM and PARAMICS. Comparison by TfL [30] showed that these two packages were generally similar, with both requiring large amounts of input data. PARAMICS is more suited to bigger networks and motorways, while VISSIM is more suited for detailed urban driving conditions. Linking the agent-based learning model with these microsimulation softwares will enhance the softwares' modelling capabilities. We propose PARAMICS as the traffic microsimulation software considering its capability to represent an individual vehicle as individual entity as well as its customizability through SNMP controllers. Our previous experiences in employing this traffic simulation software (in specific S-PARAMICS, not Q-PARAMICS) will be an advantage for this research. Developing the multi-agent learning model this way will give advantage to the current users of the corresponding microsimulation software (S-PARAMICS) with the enhancement of its capability to incorporate travellers' learning process. We aim to develop the agent-based models that have the flexibility to be customised and combined with other microsimulation softwares to give non-S-PARAMICS users the same advantage.

4 Contribution of the Research

Halcrow's [15] method for estimating cost of congestion uses presumption that the costs are calculated by comparing the time spent in a queue with the time spent if there were no congestion. This was done by their modelling work which was based on a 'fixed trip matrix', with no changes to the numbers of trips, or their frequency, or the pattern of origins or destinations, or the method of transport used, or the time of day of the journey. The only form of adaptation allowed was that instead of staying in the queue, a proportion of the drivers was allowed to travel on an alternative, albeit longer, route. This is important, as if drivers are able to adapt their behaviour in a wider range of ways, a proportion of them will find alternatives which reduce the cost of the delay (e.g. by travelling earlier or later, or putting off some discretionary journeys until next week or month, or taking the train or going to a different shopping centre). Goodwin [14] also stated that there has been little direct research on the specific nature of drivers' responses to road works, though wider research suggests that the more publicity and notice are given to works, or the longer they last, the greater adaptation is possible.

This study will fill this research gap and will eventually have impact in improving congestion prediction and with the same importance in improving one of steps in calculating costs of congestion. Microsimulation softwares have been widely used to predict the occurrences of traffic congestion and to estimate the impacts (e.g. delays) of congestion on the users of the transport system, which is also a step in calculating the cost of congestion to the economy. The improvement of their 'prediction' capability with the consideration of travellers' behavioural responses and learning process will save general public's time and therefore money as the occurrences of traffic congestion cost a lot to economy.

5 Discussion and Conclusion

The research project will combine empirical studies, which include case studies and panel survey, with experimental studies using laboratory travel choice experiment and agent-based microsimulation models. The empirical studies seek to understand the effects of traffic intervention/interruption on travellers' choices and whether and how personal experience, social information and official information influence travellers' learning and decision-making. The data obtained in the travel surveys and laboratory experiments will be used for deriving parameters as well as behavioural mechanisms in the next phase of research, developing agent-based learning models. Results of simulation are then validated with empirical data before it can be used to predict the changes of traveller's behaviour due to different scenarios implemented in the simulation experiments.

There are several challenges that may be faced during the development and application of the conceptual model. Firstly, availability of traffic data in Indonesia is a challenge as junctions or roads are mostly not automatically monitored by the traffic agency/authority. Therefore, it may require manual traffic data collection. Secondly, linking an agent-based model with a commercial microsimulation model is challenging considering the requirement of modelling the evolution of knowledge or behaviour of each individual entity. This will require researchers' advanced programming capabilities. Finally, access to a commercial microsimulation model would also be a practical challenge but could be overcome through cooperation with other institutions.

Therefore, it is expected that the aim of the study which is to better understand the evolution of traffic patterns over time, by understanding the evolution processes of knowledge which cause them, could be achieved. The research is also expected to advance knowledge about learning processes of travellers' behavioural choices and the ways to accelerate the processes and to produce a modelling technique which enables us to simulate travellers in micro level details with the capability of modelling evolution of their behaviour. The outcome of the research will enhance the capability of microsimulation softwares to accommodate travellers' learning processes and enable a better congestion prediction.

References

1. Abdel-Aty, M., & Abdalla, F. (2006). Examination of multiple mode/route choice paradigms under ATIS. *IEEE Transactions on Intelligent Transportation Systems, 7*(3), 332–348.
2. Adler, J., & Blue, V. (2002). A cooperative multi-agent transportation management and route guidance system. *Transportation Research Part C: Emerging Technologies, 10*, 433–454.
3. Arentze, T., & Timmermans, H. (2005). Modelling learning and adaptation in transportation context. *Transportmetrica, 1*(1), 13–22.
4. Balmer, M. (2007). Travel demand modeling for multi-agent traffic simulations: Algorithms and systems. Dissertation, ETH Zurich, Switzerland.
5. Bazzan, A. L., & Klugl, F. (2005). Case studies on the Braess Paradox: Simulating route recommendation and learning in abstract and microscopic models. *Transportation Research Part C: Emerging Technologies, 13*, 299–319.
6. Ben-Akiva, M., De Palma, A., & Kaysi, I. (1991). Dynamic network models and driver information system. *Transportation Research Part A, 25*, 251–266.
7. Boulter, P., & McCrae, I. (2007). *Assessment and Reliability of Transport Emission Models and Inventory Systems – Final Report. ARTEMIS deliverable no. 15*. Wokingham: TRL.
8. Dia, H., & Panwai, S. (2007). Modelling drivers' compliance and route choice behaviour in response to travel information. *Nonlinear Dynamics, 49*, 493–509.
9. Dia, H. (2002). An agent-based approach to modelling driver route choice behaviour under the influence of real-time information. *Transportation Research Part C, 10*, 331–349.
10. Fried, M., Havens, J., & Thall, M. (1977). Travel behaviour: A synthesized theory (Laboratory of psychosocial studies) Boston College.
11. Frith, B. (1999). The estimation of recurrent congestion and congestion due to roadworks and incidents: 1995/96 to 1998/99. Transport Research Laboratory Report, Crowthrone.
12. Gärling, T., Karlsson, N., Romanus, J., & Selart, M. (1997). Influences of the past on choices of the future. In R. Ranyard, W. R. Crozier, & O. Svenson (Eds.), *Decision making: Cognitive models and explanations*. London: Routledge.
13. Goodwin, P. (2004) The economic costs of road traffic congestion. Discussion paper published by Rail Freight Group. Available at: http://eprints.ucl.ac.uk/1259/
14. Goodwin, P. (2005). Utilities' street works and the cost of traffic congestion. Report available at: http://www.njug.org.uk/publication/93
15. Halcrow Group Ltd. (2004). *Assessing the extent of street works and monitoring the effectiveness of section 74 in reducing disruption, 3rd annual report, volume 3 – Estimation of the cost of delay from utilities street works, September*. London: Department for Transport.
16. Halle, S., & Chaib-draa, B. (2005). A collaborative driving system based on multiagent modeling and simulations. *Transportation Research Part C: Emerging Technologies, 13*, 320–345.
17. Hernandez, J., Ossowski, S., & Garcia-Serrano, A. (2002). Multiagent architectures for intelligent traffic management systems. *Transportation Research Part C: Emerging Technologies, 10*, 473–506.

18. Kurumatani, K. (2004). Mass user support by social coordination among citizens in a real environment. In K. Kurumatani, S. H. Chen, & A. Ohuchi (Eds.), *Multi-agent for mass user support* (pp. 1–17). Berlin: Springer.
19. Liu, R., Van Vliet, D., & Watling, D. (2006). Microsimulation models incorporating both demand and supply dynamics. *Transportation Research Part A, 40*, 125–150.
20. Logi, F., & Ritchie, S. (2002). A multi-agent architecture for cooperative inter-jurisdictional traffic congestion management. *Transportation Research Part C: Emerging Technologies, 10*, 507–527.
21. Lomuscio, A. (1999). Intelligent agents – Introduction. *Crossroads, 5*(4), 2.
22. Mahmassani, H. S., & Jou, R. C. (2000). Transferring insights into commuter behaviour dynamics from laboratory experiments to field surveys. *Transportation Research Part A, 34*, 243–260.
23. McElreath, R., Lubell, M., Richerson, P. J., Waring, T. M., Baum, W., Edsten, E., Efferson, C., & Paciotti, B. (2005). Applying evolutionary models to the laboratory study of social learning. *Evolution and Human Behavior, 26*, 483–508.
24. O'Sullivan, D., & Haklay, M. (2000). Agent-based models and individualism: Is the world agent-based? *Environment & Planning A, 32*, 1409–1425.
25. Paez, A., & Scott, D. M. (2007). Social influence on travel behavior: A simulation example of decision to telecommute. *Environment & Planning A, 39*, 647–665.
26. Polydoropoulou, A., et al. (1996). Modeling revealed and stated en-route travel response to advanced traveler information systems. *Transprotation Research Record, 1537*, 38.
27. Rosetti, R., & Liu, R. (2005). An agent-based approach to assess drivers' interaction with pre-trip information systems. *Intelligent Transportation Systems, 9*(1), 1–10.
28. Srinivasan, K., & Mahmassani, H. S. (2000). Modelling inertia and compliance mechanisms in route choice behavior under real-time information. *Transportation Research Record, 1725*, 45–53.
29. Sunitiyoso, Y., Avineri, E., & Chatterjee, K. (2009). Complexity and travel behaviour: A multi-agent simulation for investigating the influence of social aspects on travellers' compliance with a demand management measure. In E. Silva, N. Karadimitriou, & G. de Roo (Eds.), *A planners meeting with complexity*. Aldershot: Ashgate Publishers Ltd.
30. Transport for London (2003). Micro-simulation modelling guidance note for TfL.
31. Van Katwijk, R., & Van Koningsbruggen, P. (2002). Coordination of traffic management instruments using agent technology. *Transportation Research Part C: Emerging Technologies, 10*, 455–471.
32. Wahle, J., Bazzan, A. L. C., Klugl, F., & Schreckenberg, M. (2002). The impact of real time information in a two-route scenario using agent-based simulation. *Transportation Research Part C: Emerging Technologies, 10*, 399–417.
33. Yamashita, T., Kurumatani, K., Izumi, K., & Nakashima, H. (2005). *Smooth traffic flow with a cooperative car navigation system*. Paper presented in the 4th joint conference on autonomous agent and multi agent system, Utrecht.

Multi-agent Reinforcement Learning for Collaborative Transportation Management (CTM)

Liane Okdinawati, Togar M. Simatupang, and Yos Sunitiyoso

Abstract Collaborative Transportation Management (CTM) is a model collaboration in transportation area conducted through information and resources sharing. Planning and implementing CTM not only involve optimization of decisions for all collaborative agents but also involve the influence of different interactions among agents to achieve higher CTM benefits. This paper explores how agent-based modelling is used to model interaction and learning process in CTM in real systems. Agent-based model is used in this paper based on consideration that agent-based model can model the emergent decision patterns and unexpected changes of decision based on the decision-making structure. Model-free reinforcement learning is used to predict the consequences and optimize all agents' action in CTM.

Keywords Multi-agent system • Agent-based modelling • Reinforcement learning • Collaborative Transportation Management (CTM)

1 Introduction

Collaborative Transportation Management (CTM) is a form of collaboration in the transportation area. Several benefits could be perceived by CTM which are reducing the travelling time, increasing load capacity usage and minimizing transportation costs through information and resources sharing [8]. In addition, CTM also provides access to new market and knowledge sharing [4]. Previous literatures developed CTM based on the main objective to increase overall performance and agility of collaborative agents by using information sharing and resources sharing in their collaboration mechanism.

Most analytical models in CTM focused on the optimal solution for the collaborative agents are carried out through mathematical modelling, computer simulation and empirical research. However, analytical models in CTM did not answer how distributed decision-making among collaborative agents led to increase agility by

L. Okdinawati (✉) • T.M. Simatupang • Y. Sunitiyoso
School of Business and Management, Bandung Institute of Technology, Bandung, Indonesia
e-mail: aneu.okdinawati@sbm-itb.ac.id

U.S. Putro et al. (eds.), *Agent-Based Approaches in Economics and Social Complex Systems IX*, Agent-Based Social Systems 15, DOI 10.1007/978-981-10-3662-0_10

synchronizing decisions, how to synchronize decision if each collaborative agent has different objectives and different perspectives, how information is shared among collaborative agents as a foundation to develop decision and how different agents learn to align their decisions that are best for all agents.

In this paper, a multi-agent system for CTM is formulated for vertical collaboration to coordinate the process and help decision-maker to optimize the benefit for all agents involved. Agent-based model is used to model the effects of interaction among collaborative agents, the complex network of interactions and the connections of information structure and action in each collaboration stage among agents that represent real systems. Agent-based model is also used based on the consideration that agent-based model captures the learning process from the experiences and adapts it to get better result [5].

In order to model the learning process of all agents in CTM, model-free reinforcement learning method is used. Model-free reinforcement learning is also used to model how agents in CTM learn by interacting with each other in every collaboration stage. At each collaboration stage, the agents in CTM take an action, which causes different benefits, and the agent has to maximize the benefit along the course of interaction. Model-free reinforcement learning used experiences to model the learning process without using mental map to calculate the long-run benefit [3].

The rest of the paper is organized as follows. The related literatures are reviewed in Sect. 2. The problem situation is explained in Sect. 3. The simulation process is explained in Sect. 4. The results and discussion of the scenarios that illustrate this concept in the real world are explored in Sect. 5. The conclusions are given in Sect. 6.

2 Literatures

2.1 Multi-agent System

A multi-agent system is defined as autonomous entities that interact with each other by sharing the same environment [11]. In this paper, cooperative interaction is use to achieve optimal result in CTM. In multi-agent system, agents make decision based on information sharing with other agents in the environment, and each agent has different rules on how the agents respond to their decision and other agents' action [11]. The rules are either fixed or changed as the agents learn to achieve optimal result [7]. Weiß ([12]) states that multi-agent learning is defined as the interaction among agents not only to achieve a common learning objective but also to achieve different and individual learning objectives.

Weiss [11] stated that in multi-agent systems, other agents' actions influence the changes of the system environment in unpredictable ways. This makes the changes of the environment of the system uncertain and has dynamic environments [11]. Multi-agent system in this paper is used to model the relationship among agents in CTM by exchanges of information and services as well as agreements among agents. In CTM, each agent has different perspectives, different objectives and also different information that are used as the basis to make decision and action. Therefore, interactions among agents are also emphasized using multi-agent system by examining the different perspectives and influence the actions of all agents in CTM.

2.2 *Reinforcement Learning*

Reinforcement learning is a method used to capture how agent learns by interacting with its environment [9]. There are three methods of learning: (1) model-free learning, (2) model-based learning and (3) Bayesian learning [2]. This paper focuses on model-free learning, where all agents in CTM learn and adapt to maximize the benefit of joining CTM in the business network by gathering information, knowledge and feedback in each collaboration stage without using mental model. There are two learning paradigms used to optimize the benefit and action of all agents in CTM: (1) Individual learning is used as the foundation to make decision by an individual agent based on each agent's roles in collaboration stages, the rewards for each agent and their objective and perspective to join the collaboration. (2) Coordination learning is used to maintain the collaboration strategy. In this type of learning, each agent learns what is the best action and decision for action that is executed with other agents; therefore, in this learning paradigm, all agents can receive the best result if all agents involved have the best outcome with joint actions [1].

In this paper Q-learning algorithm as a popular form of model-free reinforcement learning is used to model how agents in CTM learn and adapt to choose action in each collaboration stage based on the value of each collaboration stage and used to update other agent values. Q-learning algorithm captures the interaction between the agent and environment based on trial-and-error action to obtain optimal value of the interaction [10]. In Q-learning the agent learns an action-value function or Q-learning function where an agent chooses the action which gives higher value of a previous selection of action [6]. Q-learning always selects the action that maximizes the sum of the immediate reward and the value of the immediate successor state. The Q-learning function is defined as follows:

$$Q_{t+1}(s_t, a_t) = Q_t(s_t, a_t) + \alpha_t(s_t, a_t)\left(R_{t+1} + \gamma \max Q_t(s_{t+1}, a_t) - Q_t(s_t, a_t)\right) \quad (1)$$

where $Q_{t+1}(s_t,a_t)$ = new value as Q-learning value at stage $t+1$ based on action a chosen in stage t.

$Q_t(s_t,a_t)$ = old value as Q value at stage t based on action a chosen in stage t.
α_t = learning rate.
R_{t+1} = reward action at stage $t+1$.
γ = discount factor.
max $Q_t(s_t+1,a_t)$ = estimate of optimal future value at stage $t+1$ based on action a chosen in stage t.

Learning rate ($0 < \alpha \leq 1$) represents the extent of the new information which substitutes the old information. 0 learning rate represents that the agent in the model did not learn anything, and 1 represents that the agent in the model considers only the most recent information. In this paper, $\alpha_t = 0.1$ is used as learning rate. The learning rate value strongly affects how fast the system learning converges. The higher the learning rate, the shorter the amount of time it takes to learn a good policy. However, too high a learning rate can lead to suboptimal policies. Fixed learning rate ($\alpha_t = 0.1$) is used based on the consideration that all agents in this research only used the evaluation results with equal weights in the end of agreement period to substitute the preference action in the next agreement period. Because the evaluation result is the sum of value in each collaboration stage, it means that the information used for learning rate is not only the new information; therefore, 0.1 is appropriate to represent for the learning rate.

Discount factor determines the importance of future value for the agent in the model. 0 represents that the agent only focuses on the current value, and 1 represents for a long-term high value. Discount factor $\gamma = 0$ is used based on the consideration that in CTM environment and perspective, action of other agents might be different from the previous action; therefore, the agents in CTM only focus on the current value.

3 Problem Statement

In this paper, the CTM is developed in vertical collaboration among three different agents such as a shipper, a receiver and a carrier. Information sharing mechanism is used as a foundation to make effective decisions in this collaboration. Decision-making is performed jointly by agents and implemented in the process by one agent or more agents involved.

In order to simulate collaboration process in CTM, collaboration process is divided into six stages:

(1) Formation stage – the interaction among agents is only limited to enhancing each other's attractiveness to join a collaboration.
(2) Preparation stage – each agent's interests, obligations, benefits and burdens are explored among agents.
(3) Design stage – each agent agrees on the formulation of collaboration arrangements, roles, responsibilities, etc.
(4) Planning stage – each step of decision-making for each level is defined in this stage and is used as the blueprint in the implementation stage.
(5) Implementation stage – all agents execute the planning by performing daily operations to effectively comply with the requirements of the strategic, tactical and operational objectives.
(6) Evaluation stage – evaluate the collaboration process by analysing the pay-off for each agent.

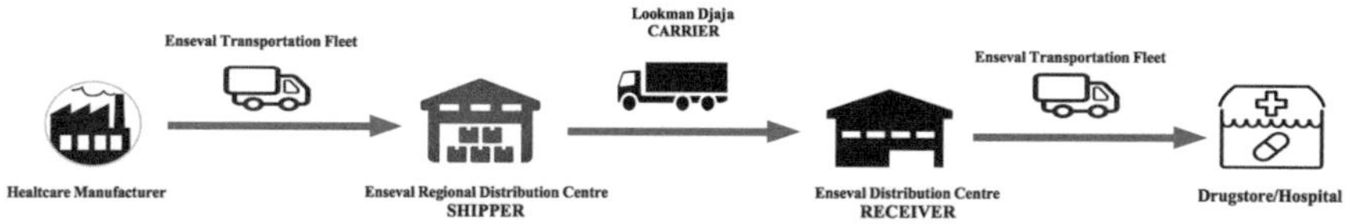

Fig. 1 Case study illustration

Case study of Lookman Djaja and Enseval used in this paper provides insights on the applicability of multi-agent model in the real system. Lookman Djaja is a transportation service provider that used a truck as their main transportation mode. Lookman Djaja's role is a carrier in CTM. Lookman Djaja provides services to Enseval to distribute healthcare products from Regional Distribution Centre (RDC) in Jakarta to Distribution Centre (DC) branch in the West area of Indonesia. Enseval has a role as shipper and receiver in CTM. The objective of Enseval is to help healthcare manufacturers to distribute their products to the customers. An illustration of the case study can be seen in Fig. 1.

Each agent in this case study has different role, perspective and objective when joining the collaboration. Learning process in CTM indicates what is the best strategy for the agents to obtain desired results in collaboration. However, due to different perspectives of agents in CTM, the actions of other agents influence the changing of rewards over time and also influence the opinion of other agents. In other words, the learning process in this paper is influenced by value paradigm.

Barrios-Aranibar and Gonçalves [1] proposed value paradigm based on the idea of action of one agent could be influenced by other agents' behaviour and action. The reward in each state is used for value function for each agent and used for evaluation of other agents on the action that is decided individually [1]. In this value paradigm, if an agent interacts with another agent and the reward of interaction is under their expectation, this result is a negative opinion for both agents.

Different agents in this case study have different criteria that count as component value. Customer value is related to quality of service, desired service that meets their needs and reliability of processes that include on-time delivery. Receiver and shipper value translated customer value as follows: (1) quality of service seen as inventory value for the shipper and receiver, (2) desired service that meets their needs seen as demand and cost value and (3) reliability of processes seen as sales and marketing value. Carrier value develops based on shipper and receiver value: (1) inventory value seen as resources, information and responsiveness value, (2) sales and marketing value seen as lead time value and (3) demand value and cost value of the shipper and receiver seen as delivery, capacity, service and cost value for the carrier.

Collaboration in this case study has not yet reached value-driven translation (stage 5 of Roddy Martin's Maturity Stage, 2012); therefore, the comparison between the actual conditions, which are functional integration and value-driven translation, will be used as scenarios to give better illustration of the benefit of CTM for all agents.

4 Simulation Process

The simulation in this paper is conducted for two scenarios: (1) the actual condition of the case study (maturity stage 3) and (2) the case study if they reach maturity stage 5. In each scenario the interaction of three agents in CTM in every stage of collaboration from formation stage until evaluation stage is captured. In each stage of collaboration, each agent has different roles and different rules. Perspectives and roles of each agent in CTM is a major concern in this paper. The interaction, roles and rules of each agent in CTM in six collaboration stages can be seen in Fig. 2.

In order to simulate the two scenarios using SOARS, the simulation process runs for 100 iterations and for 20 times. Each agent has their own criteria as component value, and each criterion has different scales of value of 0–10 and is randomly generated by computer at the initialization of the simulation process. Each agent also has different perspectives based on the component value as the agents' prior belief. Differentiation value of perspectives among agents in CTM is used in this paper as a foundation in determining behaviour, action and collaboration strategy. In addition, the positive influence of other agents is obtained if the value of one agent is above the expectation of the other agents; this causes the learning rate to become positive. On the other hand, the negative influence of other agents is obtained if the value of one agent is under the expectation of the other agents; this causes the learning rate to become negative. Therefore, there are two Q-learning functions used in this paper.

If other agents give positive influence, then

$$Q_{t+1}(s_t, a_t) = Q_t(s_t, a_t) + \alpha_t\ (s_t, a_t)\Big(R_{t+1} - Q_t(s_t, a_t) \tag{2}$$

If other agents give negative influence, then

$$Q_{t+1}(s_t, a_t) = Q_t(s_t, a_t) - \alpha_t\ (s_t, a_t)\Big(R_{t+1} - Q_t(s_t, a_t) \tag{3}$$

where $Q_{t+1}(s_t, a_t)$ = new value as Q-learning value at stage $t+1$ based on action a chosen in stage t.

$Q_t(s_t, a_t)$ = old value as Q value at stage t based on action a chosen in stage t.
α_t = learning rate.
R_{t+1} = reward action at stage $t+1$.

The summary of simulation descriptions and parameters can be seen in Table 1.

In the first stage, which is the formation stage, the value captured by three agents involved is knowledge value. The knowledge value for the shipper (RDC) and receiver (DC) is similar because the shipper and receiver are the same company – Enseval. Because the initial value in this stage generated randomly, the shipper/receiver and carrier (LD) have different values of knowledge at the first time.

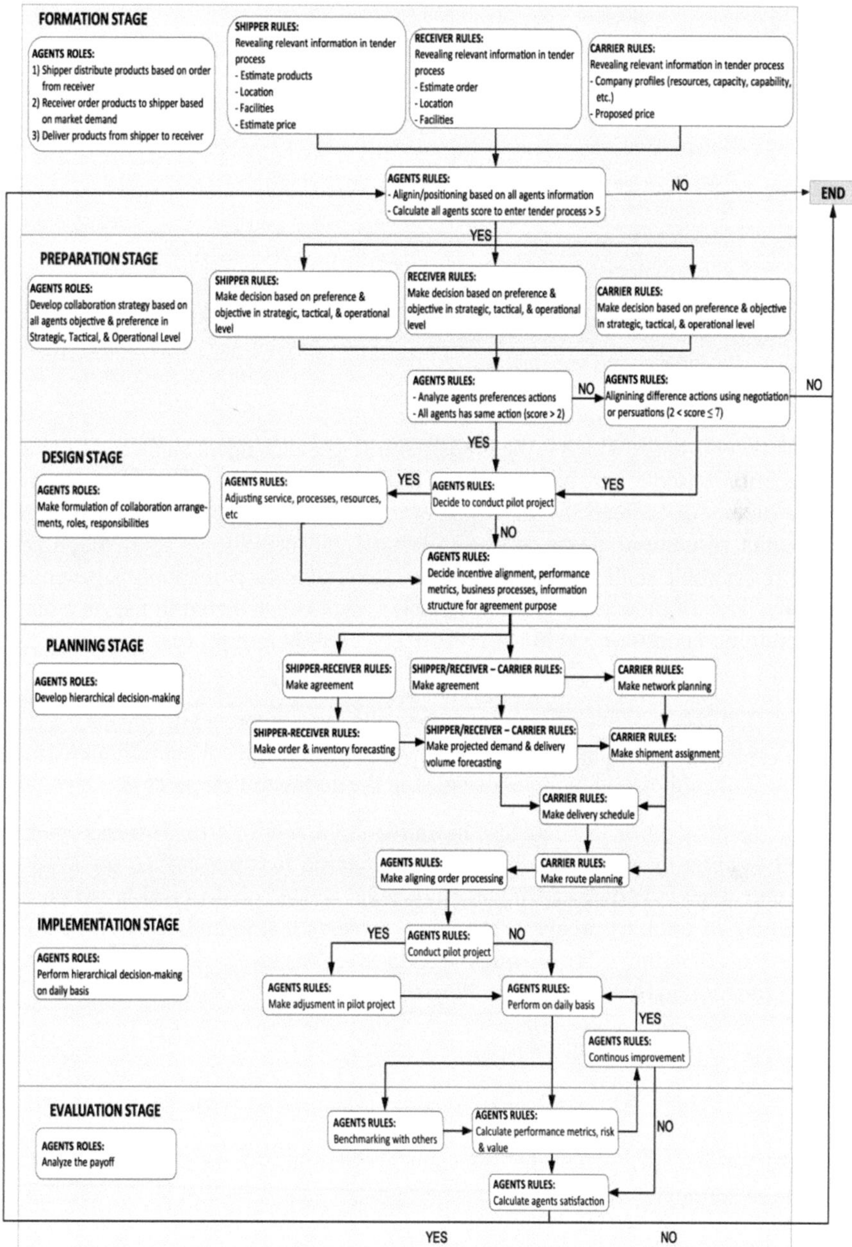

Fig. 2 Roles and rules of agents in CTM

Table 1 Summary of simulation descriptions and parameters

	Simulation scenario 1	Simulation scenario 2
Description	The actual condition of the case study, which is under maturity stage 3	The ideal condition, which is assumed if the case study reaches maturity stage 5
Agents	Shipper (Enseval Regional Distribution Centre)	
	Carrier (Lookman Djaja)	
	Receiver (Enseval Distribution Centre)	
Parameters	$\alpha_t = 0.1$ as learning rate	
	$\gamma = 0$ as discount factor	
	Value scale from 0 to 10	
	Initial value = random	
	Iteration = 100	

In the second stage, which is the preparation stage, each agent starts to deliberate its perception based on its own strategic structure formation, communication and power structure that affects the allocation of its incentive alignment, responsibilities and conflict resolutions in the strategic, tactical and operational level. Each perception has a value scale of 0–10; the initial value of each perception is generated randomly. The difference perspective among agents is calculated to use as a basic consideration to continue, negotiate or stop the collaboration process.

For this case study, $\Delta\, p$ represents the difference perspective among agents, if:

$\Delta\, p \leq 2$, then the agent will continue the collaboration process.
$2 < \Delta\, \mathrm{p} \leq 7$, then the agent will negotiate to continue the collaboration process.
$\Delta\, p \geq 7$, then the agent will choose to stop the collaboration process.

If the agents choose to negotiate, incentive alignment and performance metric are used to align the differences. The value captured by three agents involved in preparation stage is knowledge value. Knowledge value is calculated based on the learning rate of joint action algorithm that represents that knowledge value in this stage improved if the value of other agents also improved. The formula of Q-learning function in this stage is as follows:

$$Q_{t+1}\,(s_t, a_{1t}, \ldots.., a_{Nt}) = Q_t\,(s_t, a_{1t}, \ldots., a_{Nt}) + \alpha_t\,(s_t, a_t)\,.\,(R_{t+1} - -Q_t\,(s_t, a_{1t}, \ldots., a_{Nt})) \tag{4}$$

where $Q_{t+1}(s_t,a_{1t}, \ldots..,a_{Nt})$ = the new value as Q-learning value at stage $t+1$ based on the joint action a chosen in stage t by N agents.

N = number of agents.
$Q_t(s_t,a_{1t}, \ldots., a_{Nt})$ = old value as Q value at stage t based on the joint action a chosen in stage t by N agents.
α_t = learning rate.
R_{t+1} = reward action at stage $t+1$.

In design stage, the third collaboration stage, the agents in CTM make agreements that arrange roles, responsibilities, incentive alignment, decision-making structure, information structure, process integration and performance metrics of each agent. There are two values which can be captured as benefit in this stage: knowledge value and service value. Knowledge value in this stage represented increasing knowledge of each agent that learns together using coordination learning (Eq. 4) among agents in CTM when they make an agreement.

In planning stage, each step of decision-making for strategic, tactical and operational level is defined in this stage and is used as the blueprint in the implementation stage. In strategic level, each agent captures trust value, because the agreements established among agents give a reassurance for all agents and customers in which each agent promises to collaborate in accordance with the agreement and deserves to get equal benefits. The trust value for RDC and DC has the same value because the shipper and receiver are the same company. In tactical and operational level, several planning are developed from order planning until routing planning. This is done in order to make goods available and manageable to be delivered on time. Reliability value is captured both in tactical and operational level.

In the fifth stage, which is the implementation stage, all agents execute the planning by performing daily operations to effectively comply with the requirements of the strategic, tactical and operational objectives. There are three values captured by agents and customers in this stage: reliability value, service value and quality value. Even if RDC and DC are the same company, for this stage both agents calculate the value captured as benefit for each agent.

In the last stage, the evaluation process is conducted by analysing the quality and service in the end of agreement period. This becomes the input for the agents to determine whether to modify or to terminate the agreements. Service value and quality value of all agents and customers are calculated.

5 Result and Discussion

In the first stage, shipper/receiver and carrier possess the same range of knowledge value for both scenarios. This happened because all agents have learned to change their action from ending the collaboration process to continue the collaboration process to achieve better result for all agents involved. In the second stage, there are differences in the results of two scenario simulations; it can be seen in Fig. 3. Scenario 1 which is the actual condition requires a longer time to generate increased knowledge value. However, in the end of the simulation, the knowledge value for both scenarios has the same range that is a maximum range of 10.

In design stage, the third collaboration stage, there is slight difference of the two scenario simulation results. Scenario 2 generates maximum knowledge value and service value faster than in scenario 1. The service value captured in scenario 2 has smaller differences between service value of the customer and service value of all agents compared to scenario 1. This happens because in scenario 2 all agents do not hesitate to share the information. The service value in design stage is represented in Fig. 4.

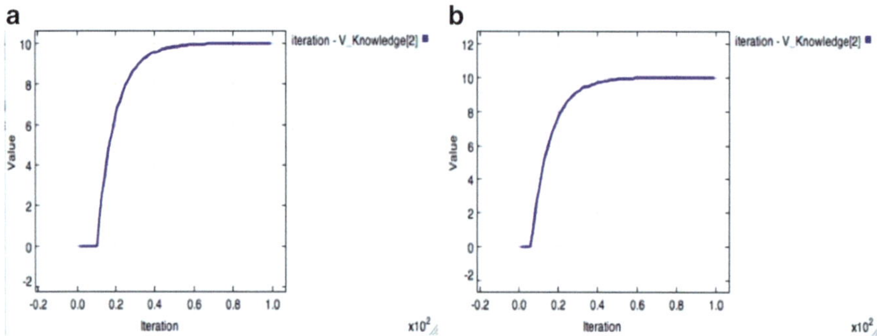

Fig. 3 Knowledge value in preparation stage: (**a**) scenario 1, (**b**) scenario 2

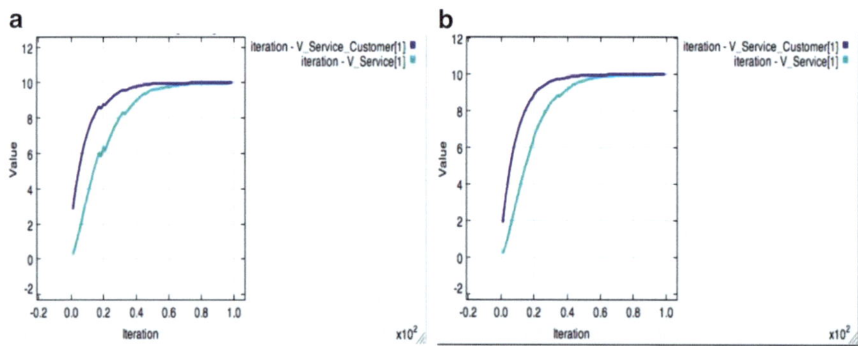

Fig. 4 Service value in design stage: (**a**) scenario 1, (**b**) scenario 2

In planning stage, both scenarios generate maximum trust value, reliability value in tactical level and reliability value in operational level at the same iteration, even though the results illustrate that the value captured in scenario 2 has smaller differences between value of the customer and value of all agents compared to scenario 1.This happens because some information such as demand/order information has not been shared with all agents in CTM in scenario 1.

In the implementation stage, because scenario 1 already integrated functional process, all agents and customers in scenario 1 are able to obtain maximum reliability value and service value similar with the simulation result of scenario 2. However, because in scenario 1 complaints only occur from one direction from the shipper and receiver to the carrier and only the small improvement do on daily basis in carrier part, quality value in scenario 1 cannot achieve a maximum range of 10 or at least the same with quality value of the customer. This also happens because in the actual condition, a pilot project is not done to gain the experience and identify the improvement for the latter's performance efficiency based on customers' value. An illustration of the simulation result in this stage can be seen in Figs. 5, 6 and 7.

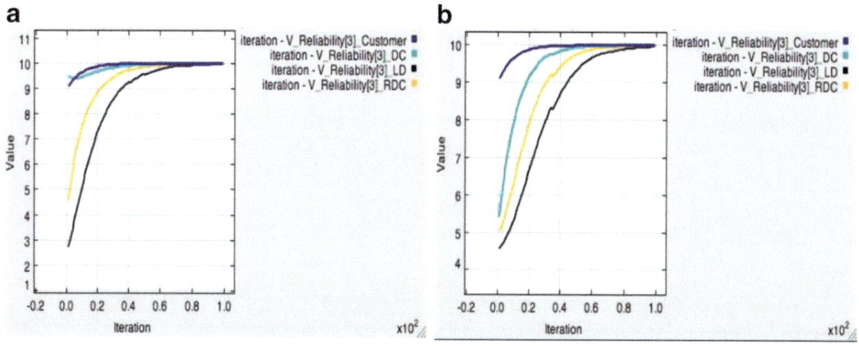

Fig. 5 Reliability value in implementation stage: (**a**) scenario 1, (**b**) scenario 2

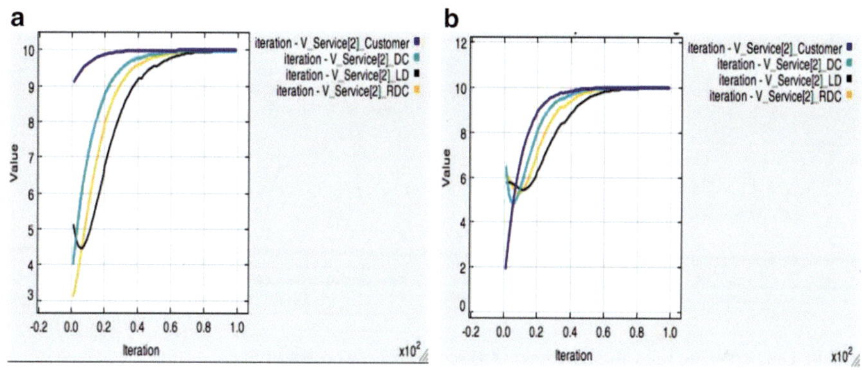

Fig. 6 Service value in implementation stage: (**a**) scenario 1, (**b**) scenario 2

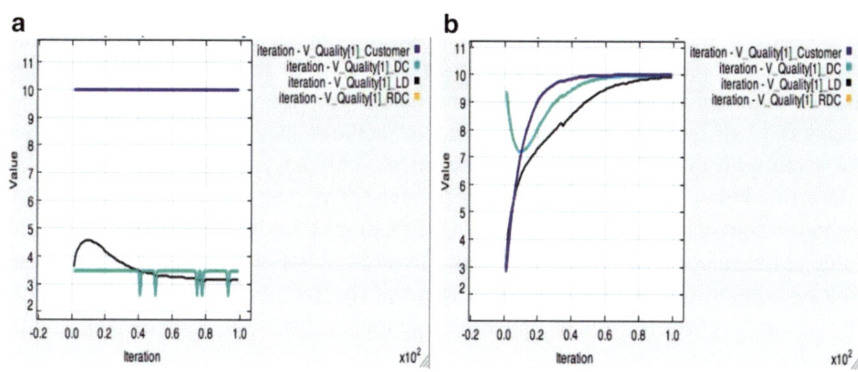

Fig. 7 Quality value in implementation stage: (**a**) scenario 1, (**b**) scenario 2

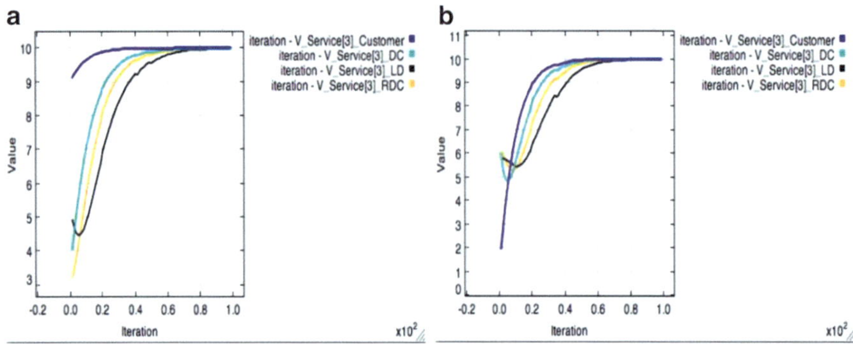

Fig. 8 Service value in evaluation stage: (**a**) scenario 1, (**b**) scenario 2

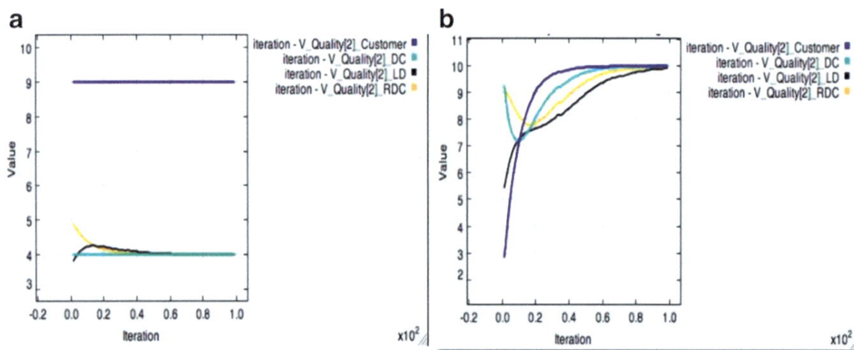

Fig. 9 Quality value in evaluation stage: (**a**) scenario 1, (**b**) scenario 2

In the evaluation process, in Fig. 8, both scenarios achieved the same value service and achieved maximum value at the same time. However, the service value in scenario 2 has smaller differences between value of the customer and value of all agents compared to scenario 1. In Fig. 9, only in scenario 2, all agents and customers can achieve the same and maximum value. On the other hand, in scenario 1, all agents and customers cannot achieve maximum range or at least the same with quality value of the customers. This happens because there is no interaction between customers and all agents in CTM. The improvement to modify the agreement and collaboration processes is also only based on the perspective and value of the shipper, receiver and carrier. Therefore, because there is no input to use as a foundation to learn, to modify or to adapt the action and value, it is difficult and seems impossible to reach maximum quality value or at least the same with customers' value in scenario 1.

6 Conclusions

Model-free learning is used in this paper to model the collaboration process in CTM, where all agents learn and adapt to maximize the benefit of joining collaboration based on the value captured of each collaboration stage and also used to update other agent values. The collaboration develops in the business network by gathering information, knowledge and feedback in each collaboration stage without using mental model. This type of learning is inclusive to trial-and-error learning; therefore, model-free learning is relatively slow to change appropriately.

In this paper, Q-learning algorithm using the influence of value reinforcement learning paradigm is used to model the learning process from the experiences and adapt it to get better result of collaboration in Lookman Djaja and Enseval case study (maturity stage 3 as scenario 1) and compare it with maturity stage 5 as scenario 2. Both scenarios are used to illustrate how interaction, learning and adaptation appear in the real system. This paper is the preliminary work of CTM in the real system. More empirical research should be carried out to determine the extent to which a multi-agent model for CTM can be implemented in the real system to capture the interaction among agents in CTM, and it is also used to explore value co-creation in CTM.

References

1. Barrios-Aranibar, D., & Gonçalves, L. M. (2009). *Influence value Q-learning: A reinforcement learning algorithm for multi agent systems*. ISBN 978–3–902613-55-4, I-Tech, Vienna, Austria, 2009.
2. Borst, M. (2013). Comparison of information theory based and standard methods for exploration in reinforcement learning, thesis Berlin University.
3. Dayan, P., & Niv, Y. (2008). Reinforcement learning: The good, the bad and the ugly. *Current Opinion in Neurobiology, 18*(2), 185–196.
4. Johnson, R. E., & Johnson, T. E. (1999). International market development through networks: The case of the Ayshire knitwear sector. *International Journal of Entrepreneurial Behaviour Research, 5*(6), 297–312.
5. Macal, Charles M., & Michael J. North. (2011). Introductory tutorial: Agent-based modeling and simulation. In *Simulation conference (WSC), Proceedings of the 2011 winter* (pp. 1451–1464), IEEE.
6. Pandey, P., Khumar, S., & Pandey, D. (2010,August). Reinforcement learning by comparing immediate reward. *International Journal of Computer Science and Information Security, 8*(5), 210–214.
7. Siebers, P. O., & Aickelin, U. (2008). Introduction to multi-agent simulation. In *arXiv preprint arXiv:0803.3905*.
8. Sutherland, J.L. (2003). Collaborative transportation management – Creating value trough increased transportation efficiencies. *Business Briefing – Pharmagenerics*.
9. Sutton, R. S., & Barto, A. G. (1998). *Reinforcement learning: An introduction*. Cambridge, MA: MIT Press.

10. Watkins, C. J. C. H., & Dayan, P. (1992). Technical note Q,-learning. *Scotland Machine Learning, 8*, 279–292.
11. Weiss, G. (1999). *Multiagent systems: A modern approach to distributed artificial intelligence*. MIT Press.
12. Weiß, G. (1996). Adaptation and Learning in multi-agent systems: Some remarks and a bibliography. In G. Weiß & S. Sen (Eds.), *IJCAI-WS 1995. LNCS (LNAI)* (Vol. 1042, pp. 1–21). Heidelberg: Springer.

Size Effects in Agent-Based Macroeconomic Models: An Initial Investigation

Shu-Heng Chen, Ying-Fang Kao, Bin-Tzong Chie, Timo Meyer, and Ragupathy Venkatachalam

Abstract We investigate the scale-free property of an agent-based macroeconomic model initially proposed by Wright (Physica A, 346:589–620, 2005), called the Social Architecture (SA) model. The SA model has been shown to be able to replicate a number of important features of a macroeconomy, such as patterns concerning economic growth, business cycles, industrial dynamics, and income distribution. We explore whether macroeconomic stylized features resulting from this model are robust when the number of agents populating the (model) economy varies. We simulate the model by systematically varying the agent population with 100, 500, 1,000, 2,000, 4,000, 8,000, and 10,000 agents. Our results indicate that the SA model does exhibit significant size effects for several important variables.

Keywords Maximum entropy principle • Size effect • Agent-based macroeconomic model • Circular flow

1 Motivation and Introduction

> One of the greatest potential contributions that ACE could make to macroeconomic theory is permitting the constructive exploration of scale effects without the external imposition of artificial coordination devices. What does it matter if an economy has 10,000 versus

S.-H. Chen (✉) • Y.-F. Kao
Department of Economics, AI-ECON Research Center, National Chengchi University, 11605, Taipei, Taiwan
e-mail: chen.shuheng@gmail.com

B.-T. Chie
Department of Industrial Economics, Tamkang University, Tamshui, 251, New Taipei City, Taiwan

T. Meyer
Department of Physics, Georg-August University Göttingen, 37073, Göttingen, Germany

R. Venkatachalam
Department of Economics, AI-ECON Research Center, National Chengchi University, 11605, Taipei, Taiwan

Institute of Management Studies, Goldsmiths, University of London, New Cross, SE14 6NW, London, UK

U.S. Putro et al. (eds.), *Agent-Based Approaches in Economics and Social Complex Systems IX*, Agent-Based Social Systems 15, DOI 10.1007/978-981-10-3662-0_11

> 300 million participants? What macroeconomic purposes are served by small-scale models, and which require a scale closer to empirical reality? *Do macroeconomies exhibit important regularities that simply cannot be generated using small scale models?* ([10], p. 248; Italics added.)

Among the many agent-based macroeconomic models, Ian Wright's *Social Architecture (SA)* model [9, 13, 14] is unique in its model-design principle. Wright models the circular flow of the macroeconomy in an agent-based fashion, i.e., by individualizing each economic action that contributes to the flow and then randomizing each action using the *entropy-maximization principle* (the EM principle).[1] With this design, no further behavioral considerations are given to the consumers, firms, employees, and employers. This is in accordance with the EM principle, which argues for allowing a maximum amount of uncertainty, given current knowledge (or the lack of it) regarding a phenomenon. While the EM principle has been applied in other agent-based economic models, such as double auction markets [7, 12] and financial markets [4, 6], to the best of our knowledge, there is no application of this principle in the literature on agent-based macroeconomic models, with the exception of Wright's model.[2]

Economists often have reservations concerning the use of the EM principle. However, the fundamental concern for the proponents of the EM principle is not whether agents behave randomly in reality, but rather the degree of empirical relevance of the given model. On this point, Wright makes a strong case for the SA model by showing that it is able to replicate several stylized facts, specifically on the distribution of a number of key economic variables, such as firms' size and demise, the duration of recessions, and the distribution of income and wealth. Perhaps the most impressive feature of this model is that it comprises just three parameters.

Given its "initial success," we would like to examine whether this model is *size-free*. Here, the "size" of the model refers to the number of agents in the model. According to [13, 14], the number of agents is merely a scaling parameter, and hence it should not affect the relative dynamics since the computational rules do not refer to them. This parameter is fixed at 1,000 throughout Wright's entire simulation. It is legitimate to wonder whether the fundamental results obtained in his analysis, such as the Zipf law concerning the firm size, would be invariant to changes in the number of agents. We ask the following questions: Are the stable patterns (distributions) discussed by Wright so *fundamental* that size plays no role in the model? If size does matter, then what properties are sensitive to size?

The significance of this study is threefold. First, size, as characterized by the number of agents, has been shown to be a crucial parameter in many agent-based

[1]The EM principle is also known as the *zero-intelligence agent* in the literature. However, as [3] has argued, this term may be misleading since the behavioral assumption may have nothing to do with the cognitive abilities of agents. Therefore, we prefer using a different and a more formal term.

[2]For a survey of the use of the entropy-maximization principle in the agent-based modeling of economics and finance, the interested reader is referred to [8].

models [2, 5, 11], where finite size effects are observed. Therefore, the results obtained using a fixed size should be carefully evaluated if they are not size-free. Second, in the context of agent-based models, it is also interesting to ask when and why size matters. Normally, size can matter because the aggregation is not merely a linear scaling-up in most agent-based settings. Interactions can alter some economic relations when size changes. Hence, by studying the size effect, we can also gain further insights into the interaction schemes employed in the model. Third, size may also matter in the real-world: For instance, the behavior of small economies can be fundamentally different from that of large economies due to size sensitivity. If so, it is desirable to identify stable, scale-independent macroeconomic properties, which can help facilitate meaningful discussions about experimental policy interventions in agent-based environments. More importantly, they can help to choose a canonical model from different candidate explanations.

The rest of the paper is organized as follows. Section 2 provides a brief overview of the Wright model, mainly, [13]. Section 3 articulates the design of the simulation, including the time structure (the time flow) of the simulation run in this paper and the parameters applied in this paper. Section 4 presents the simulation results[3] as well as an examination of the size effect, followed by a discussion and conclusion in Sect. 5.

2 An Overview of the Wright Model

2.1 *The Modeling Principle*

In the Wright model, all micro decisions are essentially random, and agents' behavior is modeled based on the EM principle. The model can be regarded as an extension of the Gode-Sunder model in the double auction market [7] to a *circular-flow macroeconomic model*. In the Gode-Sunder model, each agent needs to make only one decision at a time, either to bid or to ask, and the EM principle is applied to this single decision. Whereas in the circular-flow model agents need to make a few more decisions along the flow, and the EM principle is applied to *all* these decisions.

Figure 1 outlines a simple circular flow. There are two types of agents, households and firms, and there are two markets, a labor market and a goods market. There is neither a public sector nor a financial sector; hence the financial market does not exist in this simple circular flow. The household sector is composed of N agents (N single-head households). These N agents also constitute the labor market as employers (firms, the single-owner firms), employees, and some who are

[3]The simulations are performed using both *MATLAB* and *NetLogo*, and the codes are publicly available on the Internet (see Sect. 3).

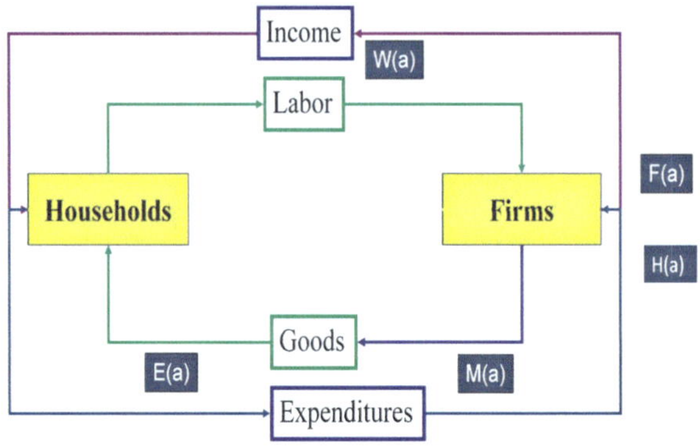

Fig. 1 Circular flow of the economy

unemployed. The products will be sold to households in the goods market. There are basically two flows that run in the reverse direction. The inside circle represents the flow of labor, products, and real consumption, and the outside circle depicts the flow of wage payments, expenditures, and firm revenues.

The circular flow can be completed in many different ways. Here, we follow Wright's sequence, i.e.,

$$\text{goods market} \rightarrow \text{labor market} \rightarrow \text{goods market}.$$

We begin with the goods market, and there is money flowing from households (expenditures) to firms (revenues). We then end up with the labor market, and there is a flow of money from firms (revenues) to households (wages). To be brief, we call this sequence $GM \rightarrow LM \rightarrow GM$.

However, from the flow as shown above, it is not clear who can buy from whom and who can hire whom. For this purpose, we incorporate a labor market mechanism that matches at the beginning of the flow, i.e.,

$$\text{L Matches} \rightarrow \text{GM} \rightarrow \text{LM} \rightarrow \text{GM}.$$

The decisions involved over the entire flow include where to work, whether to start (close) a firm, how much to consume, where to consume, how many employees to hire (fire), whom to hire, and what wages to pay. There are a maximum of nine decisions in one single flow. Of course, not all of them will be encountered in a single run; different agents may encounter different decision problems, depending on each agent's status as an employer, employee, or simply unemployed. The upshot is that, except for the closing, the firing, and the hiring decisions that are deterministic and which are dependent on the firm's working capital, the other six decisions are all based on the EM principle.

2.2 Agents and Network

We shall begin with a labor market network. We follow the notations frequently used in the network literature. Let (i,j) be a link connecting agents i and j, indicating that agent i is hired by agent j. Since (i,j) is not the same as (j,i), the labor market network considered here is *directed*. Let $g(t)$ be the labor market network at time t, which is a collection of all employee-employer relations:

$$g(t) = \{(i,j) : i,j \in \mathbf{N}, i \neq j\}, \tag{1}$$

where $\mathbf{N}$ is the set of all agents, and the cardinality of $\mathbf{N}$ is N.

Then the agent i who has a connection in the labor market can be either an employee or an employer and agent i who has no connection in the labor market is considered to be unemployed. Equation (1) already excludes the case of being self-employed, but we further assume that each agent can have at most one job, i.e., one employer. We also exclude the possibility each agent can be an employee and an employer simultaneously; hence, for agent i,

$$\text{if } \exists j, (j,i) \in g(t), \text{then } \nexists \quad k, \ni (i,k) \in g(t).$$

In other words, the network can be viewed as a *bipartite graph* at any given time.[4] Each employer can, however, have many employees, and the set of his employees (connections) at time t, $g_i(t)$, is denoted by

$$g_i(t) = \{(j,i) : (j,i) \in g(t)\}. \tag{2}$$

As for agent i who is not an employer, he is either an employee, i.e., $\exists j, (i,j) \in g(t)$, or unemployed, $g_i(t) = \emptyset$. To simplify these expressions, we use $g_i(t) = \{j\}$ ($j \in g_i(t)$) if agent i is an employer at time t, similarly $g_i(t) = j$ if agent i is an employee, and $g_i(t) = 0$, if agent i is unemployed at time t. The collection of employers (firms), employees, and the unemployed at time t will be denoted by $F(t)$, $L(t)$, and $U(t)$. Clearly, $F(t) \cup L(t) \cup U(t) = \mathbf{N}$, and $F(t) \cap L(t) \cap U(t) = \emptyset$.

2.3 Behavioral Rules in the Flow of the Economy

The economy proceeds in a way that is parallel to the circular flow, shown in Fig. 1. We run this cycle for many iterations. At time t, we begin by randomly selecting an agent from the set of agents $\mathbf{N}$ and call him agent a. Agent a can be an employer

[4]However, since the status of being an employee or an employer can change over time, a *directed graph* is more convenient.

($a \in F(t-1)$), an employee ($a \in L(t-1)$), or unemployed ($a \in U(t-1)$). Regardless of his current status, agent a will go through the economy along the sequence

$$\text{L Matches} \rightarrow \text{GM} \rightarrow \text{LM}$$

and make decisions for those circumstances which fit his status.

2.3.1 Labor Market Matches

The sequence starts with the labor market matching. If $a \in U(t-1)$, then the following action applies. Agent a enters the labor market and is randomly matched to an agent j ($j \in F(t-1) \cup U(t-1), j \neq a$) with the following probability:

$$p((a,j))(t) = \frac{m_j(t-1)}{\sum_{k \in F(t-1) \cup U(t-1)/\{a\}} m_k(t-1)}, \tag{3}$$

where $m_i(t)$ is the money holding of agent i at time t. Equation (3) implies that it is more likely for the unemployed agent a to find a job in firms which are economically more successful and have accumulated more working capital (m_k). This is similar to the *preferential attachment mechanism* used to generate the scale-free network [1]. Notice that Eq. (3) does not exclude the possibility that an agent a can be recruited by another unemployed agent j. In the latter case, a new firm is formed. The search (match) is one-shot, and its success is primarily determined by the affordability of agent j, i.e., Eq. (4):

$$g_a(t) = \begin{cases} j, & \text{if } m_j > \bar{w}, \\ 0, & \text{otherwise.} \end{cases} \tag{4}$$

where $\bar{w}$ is a *reference wage* of the labor market, to be detailed in Eq. (13).

2.3.2 Goods Market: Expenditures

After the labor market matching, the flow proceeds to the goods market. An agent j randomly selected from $\mathbf{N}/\{a\}$ enters the goods market. The total consumption of agent j at time t, $C_j(t)$, is randomly determined, following a uniform distribution:

$$C_j(t) \sim U[0, m_j(t-1)] \tag{5}$$

His consumption $C_j(t)$ will be attributed to the market value of products. Let $V(t)$ be the pool of the market value which has been accumulated up to $t-1$, i.e., the market value which has not been distributed. Then

$$V(t) = V(t-1) + C_j(t). \tag{6}$$

2.3.3 Goods Market: Revenues

From the circular flow (Fig. 1), the money in the market value (expenditure) pool, $V(t)$, will be distributed to the firms. This distribution mechanism is also random and is inclined toward the big firm. First, if agent a is not unemployed ($a \notin U(t)$), revenue is uniformly sampled from the market value pool, i.e.,

$$R(t) \sim U[0, V(t)]. \tag{7}$$

Second, this revenue will be attributed to a firm in the following manner:

$$\begin{cases} m_a(t) = m_a(t-1) + R(t), & \text{if } a \in F(t), \\ m_{g_a(t)}(t) = m_{g_a(t)}(t-1) + R(t), & \text{if } a \in L(t). \end{cases} \tag{8}$$

After the money transfer, the pool of market value will also be updated accordingly:

$$V(t) \leftarrow V(t) - R(t). \tag{9}$$

2.3.4 Labor Market: Employment and Wages

As per the circular flow, after the market value has been distributed to firms as their revenues, the transfer of this revenue to labor income or wages follows. If the randomly selected actor a is an employer ($a \in F(t-1)$), then he has to make two decisions in this stage: on employment and wages. The employment decision concerns the adjustment of the labor demand. The employer a will decide the number of workers he can afford, based on the reference wage $\bar{w}$ and his current money holdings. Let $u_a(t)$ be the number of workers that employer a has to lay off. Then the demand for labor is given by Eq. (10):

$$u_a(t) = \max(\mid g_a(t-1) \mid - \lfloor \frac{m_a(t)}{\bar{w}} \rfloor, 0), \tag{10}$$

where $\mid g_a(t-1) \mid$ denotes the number of workers hired by employer a at time $t-1$; $\lfloor x \rfloor$ is the greatest integer that is less than or equal to x. $\lfloor \frac{m_a(t)}{\bar{w}} \rfloor$, therefore, is the affordable demand for labor, and subtracting it from $\mid g_a(t-1) \mid$ gives the number of layoffs. The layoffs will be randomly selected from $g_a(t-1)$. Notice that the *sup* operator, max, is imposed; hence the labor demand can only be adjusted downward, not upward. In the worst case, firm a may fire all its workers. If that happens, the status of agent a will change from that of an employer ($g_a(t-1) = \{j\}$) to unemployed ($g_a(t) = 0$). The status of all those who are fired also needs to be updated as well. Let the set $U_a(t)$ be the set of agents who are fired by firm a; then

$$g_i(t) = 0, \forall i \in U_a(t). \tag{11}$$

After the employment decision, if

$$g_a(t) = g_a(t-1)/U_a(t) \neq \emptyset,$$

then the remaining workers, $i \in g_a(t)$, will be paid their wages according to the following random mechanism:

$$w_i(t) \sim \begin{cases} U[w_L, w_H], & \text{if } m_a(t_i) \geq w_H, \\ U[0, m_a(t_i)], & \text{otherwise,} \end{cases} \tag{12}$$

where $m_a(t_i)$ is the money holdings of firm a upon the moment of paying wages to employee i, and w_L and w_H are the floor and the ceiling of the wage range for the employee. When firm a's money holdings are sufficient to cover the upper bound of this range, Eq. (12) states that the wage will be randomly determined according to the uniform distribution $U[w_L, w_H]$. The reference wage $\bar{w}$ which we use in this paper is in fact the midpoint of this range, i.e.,

$$\bar{w} = \frac{(w_L + w_H)}{2}. \tag{13}$$

Nonetheless, upon the moment of paying wages to employee i, if employer a finds that his money holdings are less than the threshold w_H, the wage will be uniformly randomly determined from 0 to the firm's maximum affordability, $m_a(t_i)$. After paying all wages to his employees, the money holdings of employer a will be updated again as follows:

$$m_a(t) \leftarrow m_a(t) - \sum_{i \in g_a(t)} w_i(t), \tag{14}$$

and in the meantime,

$$m_i(t) = m_i(t-1) + w_i(t), \forall i \in g_a(t). \tag{15}$$

This then comes to the end of time t. For all other agents i who are not involved in any part of the circular flow, either as an unemployed worker, an employer, an employee, or a consumer, his money holding will remain unchanged, namely, $m_i(t-1) = m_i(t)$. The state of the economy at time t, $S(t)$, can be summarized as a collection of triplets:

$$S(t) = \{(m_i(t), g_i(t), w_i(t))\} : 1 \leq i \leq N\} \tag{16}$$

To sum up, Wright's agent-based model of the circular flow can be perceived as a "relay race." Each "runner" (agent) is randomly sampled from **N** with replacement. When one runner, at time t, "runs through" the flow, the next runner, at time $t+1$, takes the baton and does the next run through. This cycle will be repeated for a sufficiently large number of times so that almost all agents have been introduced to the economy at least once (on average) within a given length of time, say, a day, a week, or a month. Then the clock moves to the next day, week, or month, and so on.

3 Simulation Design

In our simulation, we have running time in *ticks* (the tick data), loops of ticks (the monthly data), and loops of loops (the yearly data). Each tick t represents one iteration, which corresponds to the time interval $[t, t-1)$ and also to one round of the sequence L Matches $\rightarrow$ GM $\rightarrow$ LM. The iterations will be repeated N (i.e., the number of agents) times and this together forms one *loop* (a duration of one month) of simulations. By setting the number of iterations as N, we ensure that each agent, on average, will be sampled once *per month*. The entire simulation is run for a number of "years." The notation $X(t_{mo}^{yr})$ indicates the variable A at the tth iteration of the month mo of year yr. In this paper, $t = 1, \ldots, N$, $mo = 1, \ldots, 12$, $yr = 1, \ldots, 100$. In this setting, our time scale and duration of the simulation is entirely consistent with [13].

In this article, instead of merely replicating Wright's model, we alter the size of the population, N, for the purpose of examining the possible size effects, in general. Therefore, in some cases we do not have an a priori assumption concerning distributions as Wright would have had. For example, we do not assume a priori that the GDP growth rate follows a Laplace distribution. Instead, we examine whether the distributions are sensitive to size. We consider 100, 500, 1,000, 2,000, 4,000, 8,000, and 10,000 agents (Table 1). To focus on the size effect alone, we keep the money holding per capita fixed throughout all simulations at 100, which is independent of the size of population.[5] We repeat the simulations 100 times for each size variation. We also compare the resulting distribution of these estimates using appropriate statistical tests. Table 1 gives a summary of the values of the control parameters used to run the simulation of this paper. The results reported in this paper are all obtained from MATLAB.

Table 1 Table of control parameters

N	Number of agents	100, 500, 1,000, 2,000, 4,000, 8,000, 10,000
$\bar{m}$	Average money holding	100
w_L	Wage (lower bound)	10
w_H	Wage (upper bound)	90
Yr	Number of years per simulation	100
	Number of trials	100

[5]Notice that, since the amount of money is initially randomly distributed among all agents, we can only fix the money holding per capita, but not the possible distribution effect.

We collect monthly data on different variables during the course of the simulation and group them together in frequencies (monthly, yearly) along the same lines as [13]. The variables (data) and their different frequencies are defined as follows:

- **Monthly observations** Monthly data are collected for the following variables: (1) *firm size* (*number of employees* observed at each firm, at the end of each month) and (2) *firm demise (ratio)*,[6] the number of firms that cease to exist during each month in the economy.
- **Yearly observations** Yearly data are collected for the following variables: (1) *percentage of capitalists*, *percentage of workers*, and *unemployment rate*
  (2) *growth of firms* (size of sales and employment), (3) *GDP growth*, (4) *wage and profit share*, (5) *total wealth*, (6) *profit rate*,
- **Long-run observations** *Duration of recession*: A recession is said to begin in *yr* when the output expansion ceases and the economy begins to contract. The recession ends when the reverse happens.

Examining the distributions for each of these variables and their respective conformity with the observed stylized facts is in itself an interesting task. However, for this paper, we focus exclusively on the size effects of the SA model and not on their degree of conformity with actual stylized facts. We examine whether economies of different sizes (N) will generate different distributions for each of the macroeconomic variables. Since the underlying data generating process as specified by the entropy-maximizing model is stochastic, to obtain statistical reliability for the simulation results, for each treatment we run trials 100 times (Table 1). Therefore, for each variable, we obtain 100 ensembles, and our analysis of the size effect is then based on a comparison across these 100 ensembles.

Since we are comparing different distributions and variables with different frequencies, we need to choose a representative distribution for each treatment, so that we perform comparisons across treatments. Once we fix the relevant period that we wish to consider for each variable, we pool the data across repeated simulations (100 of them) for each treatment. For this purpose, we consider the last few periods for each variable under the assumption that their distributions stabilize over the course of the simulation.[7] Table 2 provides a summary of the variables, data type used, duration of the data, and statistical test employed. We then compare the representative distributions across treatments to examine the size effects. Although some of our results in the replications may not fully confirm with what Wright has demonstrated, especially with regard to replicating specific aggregate distributions, we believe that our simulation is faithful to his protocol. This exercise can, in its own right, be considered as an independent study that investigates the size effects in a Wright-like model.

---

[6]Since we compare distributions across different sizes of the economy, we normalize the absolute value of the number of firm demises into *ratios* with respect to the size of the economy.

[7]See [13], p. 598, for the rationale behind doing so.

Table 2 Range of data and statistical tests

Variable	Data type	Duration	Test
Capitalist ratio	Historical means	All	Wilcoxon RS
Worker ratio	Historical means	All	Wilcoxon RS
Unemployment ratio	Historical means	All	Wilcoxon RS
Firm demise ratio	Historical means	All	Wilcoxon RS
Wealth Gini	Historical means	Last 10 years	Wilcoxon RS
Yearly log GDP growth	Pooled	Last 10 years (1200 obs)	K-S
Yearly wage rate	Pooled	Last 10 years (1200 obs)	K-S
Firm growth (employment)	Pooled	Last 10 years	K-S
Firm growth (sales)	Pooled	Last 10 years	K-S
Rate of profit	Pooled	Last 10 years	K-S
Recession yearly	Pooled	All	K-S
Firm-size dist	Pooled	Last month	K-S

Wilcoxon RS Wilcoxon Rank sum test, *K-S* Kolmogorov-Smirnov test

4 Simulation Results and Discussion

From our simulation of the Wright model, we find that the resulting distributions of several macroeconomic variables seem to be *size dependent*. We use the Kolmogorov-Smirnov test, a nonparametric test, to compare different distributions across treatments. When we compare means across different distributions, we use the Wilcoxon rank sum test. The rank sum test is less sensitive to outliers compared to the two-sample t-test. Although our sample size is big enough to use the t-test, our data appear to have heterogeneous variances, and therefore the Wilcoxon rank sum test is more appropriate.

Table 3 presents the results of the statistical tests that compare the distributions for different sizes. We use 1% as the significance level, and each cell in the table reports the result of the pairwise comparison of the distribution of macroeconomic variables across different sizes, evaluated using an appropriate statistical test.[8] The null hypothesis of the Wilcoxon rank sum test is that the two samples have identical medians, and the null hypothesis of the K-S test is that the two distributions are identical. The check mark indicates that the null hypothesis is rejected for that cell, which lends support for the existence of the size effect. From Table 3, it is clear

[8]For the less stringent case of $p = 0.05$, a few combinations also exhibit size effects in addition to those in Table 3. The variables and the corresponding combinations are worker ratio (100/2,000, 100/8,000, 4,000/10,000, 8,000/10,000), firm demise ratio (1,000/2,000), wealth Gini (500/1,000), yearly wage share (8,000/10,000), firm growth (employment) (500/1,000, 500/2,000), rate of profit (100/500, 100/2,000, 500/2,000, 2,000/4,000), and recession yearly (4,000/10,000). However, the overall patterns concerning the size effects remain unchanged.

Table 3 Significance results for size effects - p=0.01

	100						500					1,000				2,000			4,000		8,000
	500	1,000	2,000	4,000	8,000	10,000	1,000	2,000	4,000	8,000	10,000	2,000	4,000	8,000	10,000	4,000	8,000	10,000	8,000	10,000	10,000
Capitalist ratio	√	√	√	√	√	√	√	√	√	√	√	√	√	√	√	√	√	√	√	√	
Worker ratio	√			√																	
Unemployment ratio	√	√	√	√	√	√	√	√	√	√	√	√	√	√	√	√	√	√	√	√	√
Firm demise ratio	√	√	√	√	√	√	√	√	√	√	√		√	√	√	√	√	√			
Wealth Gini	√	√	√	√	√	√		√	√	√	√		√	√	√	√	√	√	√		
Yearly log GDP growth	√	√	√	√	√	√	√	√	√	√	√		√	√	√	√	√	√	√	√	√
Yearly wage rate	√	√	√	√	√	√	√	√	√	√	√	√	√	√	√	√	√	√	√	√	
Firm growth (employment)	√	√	√	√	√	√			√	√	√			√	√	√	√	√	√	√	
Firm growth (sales)																		√		√	√
Rate of profit				√	√	√			√	√	√		√	√	√		√	√	√	√	√
Recession yearly																		√			
Firm-size dist	√	√	√	√	√	√	√	√	√	√	√	√	√	√	√	√	√	√	√	√	√

The table indicates the results of the pairwise comparisons of variables across different size treatments. The Wilcoxon Rank sum test has been used to test the size effects of the respective variables and significance levels are based on 2-tailed tests. The rest of the variables are compared using the Kolmogorov-Smirnov test. √ indicates that the distribution is significantly different between the two corresponding treatments that most variables seem to exhibit size effects. The only exceptions are the worker ratios, the duration of the recession (k), and firm growth distributions. The latter two seem to exhibit size effects for some ranges but are not consistent over all sizes.

For variables involving class ratios such as the capitalist ratio, unemployment ratio, and worker ratio, we do observe a size effect. For the capitalist ratio, there is a trend of a declining mean (and the variance of this mean) as the size of the economy increases. For the unemployment ratio, the trend is the opposite and the mean ratio increases with size. There is no discernible trend in the worker's ratio, except for the fact that they are significantly different across sizes of the economy. The firm demise ratio exhibits a roughly declining trend when the size increases from a small value to a medium one. However, with over 4,000 agents, the size effects seem to disappear. The wealth distribution is analyzed across treatments by comparing the mean Gini coefficients for the last 10 years (pooled for 100 repetitions) across different treatments. We find that the wealth distribution does exhibit size effects, but ceases to exist once the economy becomes large, indicating that the inequality tends to stabilize.

The distribution of yearly GDP growth rates evaluated using K-S tests indicates that there are significant size effects across treatments. The wage share distribution also indicates that there are significant size effects across treatments. Note that although the results of the K-S tests suggest that there are significant differences between the treatments, it is hard to conclude whether these differences are related only to the magnitude, or to the distribution, or to both.

The duration of the recession is calculated in retrospect after the simulation has been completed for 100 years. This variable by and large does not seem to exhibit size effects except for one or two cases. The firm growth distributions are examined for two different definitions of growth – in terms of sales and employment. In both cases, there are size effects, albeit for different, limited ranges. As for the employment-oriented definition, size effects are more evident. The rate of profit displays size effects, except in smaller economies. Firm-size distribution, on the other hand, exhibits very clear size effects. Figure 2 illustrates the normalized firm-size distributions for different agent size specifications.

Figure 3 demonstrates four aggregate variables that have demonstrated size effects (for the median) throughout different treatments. The medians of the capitalist and unemployment ratios exhibit a decreasing and increasing trend with an increase in the size of the population, leaving worker ratios relatively static. The firm demise ratios exhibit a mild decreasing trend, although the magnitudes are negligible. The Gini coefficients of wealth distribution also exhibit a slightly increasing trend. One noticeable feature is that the variance (or the range of the data) decreases as the population of agents increases. We find that the smaller economies tend to have higher variances for the above specified variables, and this could be because these variables are expressed in terms of ratios.

Some remarks are in order. When the K-S test does not reject the null hypothesis, we can be confident that both the distribution and the magnitude have no significance. However, in cases where the K-S test does reject the null hypothesis, we need to be more cautious about the conclusions that we draw. Consider two treatments

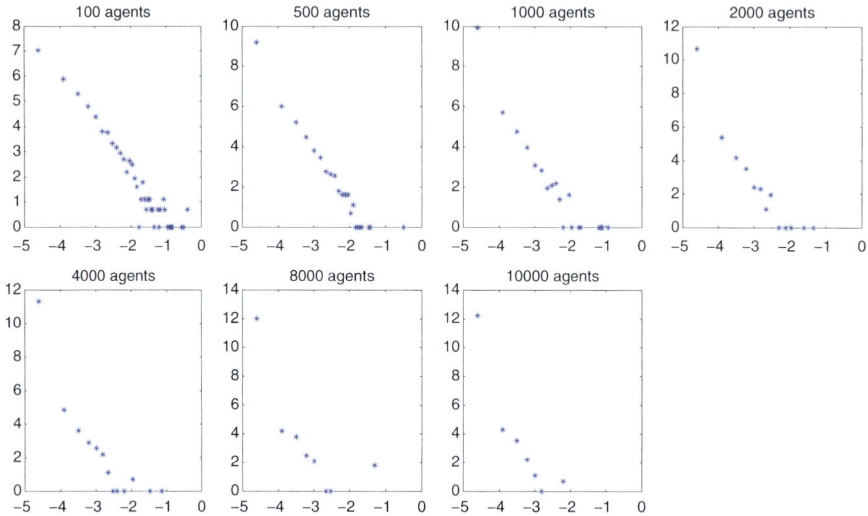

Fig. 2 Firm-size distribution (normalized): the figures demonstrate the log-log plot of firm-size distribution after normalizing it for the size of the economy since the absolute size of firms in the larger economy will be naturally bigger. The plots of smaller economies are visibly different from those of the larger economies

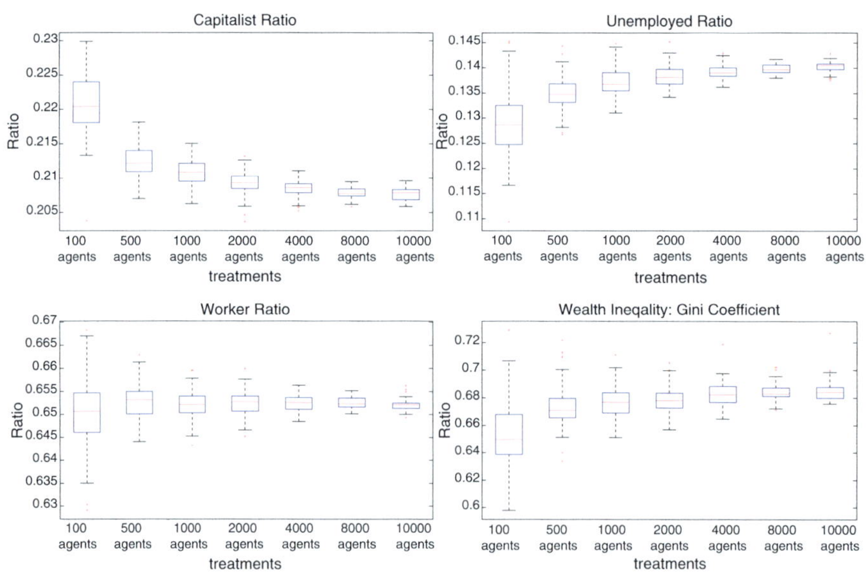

Fig. 3 Size effects of different class and macroeconomic variables (capitalist ratio, unemployment ratio, worker ratio, and wealth distribution (Gini coefficients)) are shown above. The deviations in their means across repetitions tend to become smaller as the sizes of the economies increase

resulting in different distributions, but with almost the same means. According to the K-S test, it is only suggested that one treatment makes the distribution more diverse or skewed.

5 Conclusion

We have examined the scale-free property of an agent-based macroeconomic model initially proposed by [13]. We find that many variables used in the model exhibit size effects. The issue of a size effect is more intricate than what we had expected. Some variables, especially those expressed in ratios, tend to stabilize with the systematic increase in the number of agents in our simulations, with a reduction in the spread of the means of these ratios. If such a stabilization underpins a model, then it certainly becomes important to pay attention to size specifications. It is also worth mentioning that our mode of investigation may not be the only rigorous way of testing this size effect conjecture. We have adopted this approach to remain close to the analysis performed by Wright and to do so in the most intuitive way possible. Further research is required on this front.

Although we find some initial evidence supporting size effects in this paper, further analysis is needed to unearth the possible mechanisms that make size a decisive variable. The appearance of the size effect for different variables may have unique causes for each of them, and our interest, as a first step, has been to unearth common structures which generate this possibility. A comprehensive examination of the "general" cause for the size effect requires an approach similar to "big data analysis," which is beyond the scope of this paper. Our initial investigation can be seen as a first step to highlight the potential that underlies agent-based models to examine the possible roles of size in the macroeconomy.

Acknowledgements The first and the last authors are grateful for the research support in the form of the Ministry of Science and Technology (MOST) Grants, MOST 103-2410-H-004-009-MY3 and MOST 104-2811-H-004-003, respectively.

References

1. Barabasi, A.-L., & Albert, R. (1999). Emergence of scaling in random networks. *Science, 286*, 509–512.
2. Challet, D., & Marsili, M. (2003). Criticality and market efficiency in a simple realistic model of the stock market. *Physical Review, 68*(3), 036132-1–036132-4.
3. Chen, S.-H. (2012). Varieties of agents in agent-based computational economics: A historical and an interdisciplinary perspective. *Journal of Economic Dynamics and Control, 36*(1), 1–25.
4. Duffy, J., & Unver, M. (2006). Asset price bubbles and crashes with near zero-intelligence traders. *Economic Theory, 27*, 537–563.

5. Egenter, E., Lux, T., & Stauffer, D. (1999). Finite-size effects in Monte Carlo simulations of two stock market models. *Physica A: Statistical Mechanics and Its Applications, 268*(1), 250–256.
6. Farmer, D., Patelli, P., & Zovko, I. (2005). The predictive power of zero intelligence in financial markets. *Proceedings of the National Academy of Sciences, 102*, 2254–2259.
7. Gode, D., & Sunder, S. (1993). Allocative efficiency of markets with zero intelligence traders: Market as a partial substitute for individual rationality. *Journal of Political Economy, 101*, 119–137.
8. Ladley, D. (2012). Zero intelligence in economics and finance. *Knowledge Engineering Review, 27*(2), 273–286.
9. Lavicka, H., & Novotny, J. (2013). Employment, production and consumption with random update: Non-equilibrium stationary state equations. *Acta Polytechnica, 53*(6), 847–853.
10. LeBaron, B., & Tesfatsion, L. (2008). Modeling macroeconomies as open-ended dynamic systems of interacting agents. *American Economic Review, 98*(2), 246–250.
11. Lux, T., & Schornstein, S. (2005). Genetic learning as an explanation of stylized facts of foreign exchange markets. *Journal of Mathematical Economics, 41*(1–2), 169–196.
12. Sunder, S. (2004). Market as artifact: Aggregate efficiency from zero-intelligence traders. In M. Augier & J. March (Eds.), *Models of a man: Essays in memory of Herbert A. Simon* (pp. 501–519). Cambridge: MIT Press.
13. Wright, I. (2005). The social architecture of capitalism. *Physica A, 346*, 589–620.
14. Wright, I. (2009). Implicit microfoundations for macroeconomics. *Economics: The Open-Access, Open-Assessment E-Journal, 3*, 1–27.

The Effectiveness of Firm-Controlled Supporters to Pacify Online Firestorms: A Case-Based Simulation of the "Playmobil" Customize-It Incident

Nikolaus Franke, Peter Keinz, Alfred Taudes, and Thomas Funke

Abstract We examine the effectiveness of employing firm-controlled supporters to pacify online firestorms via a case-based agent-based model of the Playmobil customize-it firestorm. We enrich traditional opinion dynamics models by integrating empirically observed aspects and use real-world data to calibrate and validate the model. Our simulations show that even a single firm-controlled supporter may significantly help pacify such a conflict. Moreover, such a firm-controlled infiltrator does not even have to be an "opinion leader" within the community in order to calm a conflict within a community.

Keywords Case-based agent model • Customer community • Firestorm

1 The "Playmobil" Customize-It Incident

For many years, "Playmobil" benefited from the activities of its customer community "customize-it," a network of creative "Playmobil" users. The company used to support "customize-it" by providing its members with free toys and materials. In turn, "Playmobil" screened the community activities in order to find ideas for new product lines. However, in spring 2009, a conflict between "Playmobil" and "customize-it" arose when Markus Bomhard, an evangelic pastor, started to adapt "Playmobil" toys for the purpose of reenacting bible scenes. "Playmobil" accused Mr. Bomhard of presenting his figures in a disgraceful and infamous manner and claimed that customizing itself involved several risks. From that moment on, both parties only communicated through their respective attorneys, and as a consequence the conflict escalated. The conflict and its implications were discussed and carried

---

N. Franke • P. Keinz (✉) • A. Taudes
WU – Vienna University of Economics and Business, Welthandelsplatz 1, 1020 Vienna, Austria
e-mail: nikolaus.franke@gmail.com; peter.keinz@wu.ac.at; alfred.taudes@wu.ac.at

T. Funke
TechQuartier, Platz der Einheit 2, 60327 Frankfurt am Main, Germany
e-mail: thomas.funke@gmail.com

U.S. Putro et al. (eds.), *Agent-Based Approaches in Economics and Social Complex Systems IX*, Agent-Based Social Systems 15, DOI 10.1007/978-981-10-3662-0_12

on within the entire community over a period of 2 months, leading to a great wave of public attention not just within the community. The conflict was incorporated into various blogs and, additionally, the engagement of traditional media added to the damage of the company's reputation and image and the demise of the "customize-it" community.

2 A Case-Based Agent Model of Online Firestorms and Infiltration Strategies

2.1 An Emerging Phenomenon: Online Firestorms

The "Playmobil" customize-it incident is an example of an online firestorm. Online firestorms can be regarded as fast-evolving cascades of negative word of mouth within online communities or social networks [26]. Clearly, the way "Playmobil" handled the case was not optimal. Another possible strategy would have been the infiltration of the community by sending out disguised firm-controlled individuals to participate in the online customer community in which the online firestorm is about to emerge. These firm-controlled supporters are masked as ordinary customers and ordered to constantly spread positive word of mouth about the company [24]. The idea is to positively influence the peer customers' attitude toward the company and thereby prevent them from participating in negative word-of-mouth cascades, thus reducing the magnitude of the online firestorm [26].

"Infiltration" has proofed as an effective strategy to resolve conflicts in many offline contexts, e.g., employing "labor spies" to prevent and/or control strikes [13] or using "demonstration marshals" to control and calm down demonstrations ([16]). However, online firestorms differ from conflicts in offline settings with regard to their highly dynamic and emergent nature as well as their level of affectivity [12, 19, 20]. Thus, it seems questionable if and under which circumstances single or a relatively small group of firm-controlled supporters will be actually capable of influencing a vast majority of peer customer community members and thereby pacify an emerging online firestorm [6, 17, 25, 26]. In this paper, we systematically investigate the effectiveness of the "infiltration" strategy. Based on data of a real-life online firestorm, we build a case-based agent-based model of a specific online firestorm capable of validly simulating the emergent dynamics of this conflict. In a second step, we conduct a number of sensitivity tests in order to analyze those factors at the level of the firm-controlled supporters and the online customer community that literature regards as most important influentials of the effectiveness of "infiltration" strategies. In particular, we analyze whether 1.) firm-controlled supporters are actually capable of effectively affecting the public opinion within the online customer community at the end of an online firestorm as well as the conflict's dynamics and whether 2.) this capability is moderated a) by the number and/or network position of firm-controlled supporters, b) the average attitude toward the company among the online customer community members prior to the conflict, and c) the proportion of the online customer community members potentially affected by the company's misbehavior. By answering these questions, we contribute to the literature on management of online complaint behavior.

2.2 Theory of Public Opinion Formation

The "theory of public opinion formation" by Katz and Lazarsfeld [14] is a good starting point when investigating the effectiveness of the infiltration strategy in a model-based way. However, traditional "opinion dynamics" models often do not account for personal characteristics of "extremists" [4, 14, 17, 31]. Also, whether members of an online customer community participate in such a conflict depends on their wish to retaliate the company for a perceived misbehavior [11, 29]. Furthermore, if customers perceive themselves [22, 23], their peers [30], or the online community of which they are members [2, 27] affected by the company's misbehavior, they will be motivated to engage in an online firestorm. We therefore developed a case-based agent-based model [3] to study the infiltration strategies against firestorms by simulation.

2.3 The Design of the Model

The model distinguishes between customers that publicly complain about a company and start a negative word-of-mouth cascade, firm-controlled supporters constantly spreading a positive word of mouth about the company, and average customers who can either participate in the online firestorm by spreading negative or positive word of mouth or stay neutral and inactive. Our model comprises a total of 363 agents. The number of agents of type one is always 1 at the outset of the conflict. The number of firm-controlled supporters is – depending on the simulated scenario – either zero, one, or ten.

Each agent is described along six different properties: the agent's initial attitude toward the company (positive/neutral/negative), the agent's confidence about his or her own opinion, the agent's authoritativeness (measured by the agent's role, e.g., "opinion leader" vs. "follower" as well as his or her betweenness centrality [9, 14]), the agent's trustworthiness (tie strength to other members; see [15]), the agent's current level of activity (only receiving word of mouth vs. receiving and transmitting word of mouth), and the perceived affectedness of oneself, the peers, and the community as a whole [30] by the company's misbehavior that has led to the publicly expressed complaint.

Two different agent behaviors are specified. The first behavior is the (passive) adoption of a specific attitude toward the company. The probability to adopt a peer's attitude depends on an agent's level of confidence about his or her own attitude as well as the perceived authoritativeness and trustworthiness of the peer sending community member [10]. An agent's level of confidence about his or her attitude toward the company itself decreases in cases when agents are confronted with diverging opinions of peers that exhibit high levels of authoritativeness and trustworthiness (i.e., if they are opinion leaders and/or connected to the agent via a strong tie). The second behavior agent in our model can show is proactive participations in the online firestorm by transmitting their attitude toward the company. If an agent participates in the online firestorm by sharing his or her negative/positive opinion about the company basically depends on the level of perceived affectedness by the misbehavior of the company. If an agent thinks that he or she, his or her peers, or the online customer community as a whole might be affected by "Playmobil's" prohibition to customize "Playmobil" toys, he or she will be more likely to change from a passive into a proactive status and engage in the negative word-of-mouth cascade.

The environment in which our agents are active is a scale-free social network [1]. In our online customer community, all postings are equally available to all members; however, messages are not necessarily read by everyone within the community. The probability to read a specific message, e.g., the original complaint or related postings sharing a negative word of mouth, depends on 1) random effects and 2) the existence of ties with the member who had posted the message.

In the initialization step of our model, the agents (with their individual properties) and the network is created. One agent is randomly selected as the original complainer who starts to spread a negative word of mouth. In addition and depending on the simulated scenario, zero, one, or ten agents are randomly assigned the role of a firm-controlled supporter and start spreading a positive word of mouth. In the subsequent iteration steps, the other agents (depending on the behaviors described above) adopt/do not adopt the negative/positive opinion toward the company and stay inactive/proactively participate in the conflict. The model terminates after the last iteration step, i.e., if no more changes in the agents' opinions occur.

2.4 Validation of the Model

In order to validate the model, we checked whether the properties and behaviors of the agents within our model correspond in a meaningful way to the properties and behaviors of the real-life actors that are to be represented by the agents (micro-face validation; see Rand and Rust [21]). We also investigated whether the macro-level patterns in our model (i.e., the diffusion of negative word of mouth) corresponds to real-world patterns (macro-face validation; see North and Macal [18]). Both, micro- and macro-face validation was done based on current literature on online firestorms [20, 26, 28] and the insights gained from six in-depth case studies of recent online firestorms, including our "Playmobil" case. We conducted 64 qualitative interviews with different parties involved in the respective online firestorms, ranging from

Table 1 Threshold assignment based on past messaging behavior

Posted messages in the month prior to the conflict	Assigned threshold
101–200	0,8 – 0,9
201–300	0,7 – 0,8
301–400	0,6 – 0,7
401–500	0,5 – 0,6
501–600	0,4 – 0,5
601–700	0,3 – 0,4
701–800	0,2 – 0,3
801–900	0,1 – 0,2
901–5000	0 – 0,1

Table 2 Coding scheme for role properties of agents (excerpt)

Name	Posts	Betweenness centrality	Opinion leadership	Role
Andy	399	2,05	No	Follower
Andyente	416	147,18	Yes	Opinion leader
Angela	22	0	No	Follower
Argento	105	8,06	No	Follower

community members and forum administrators to company representatives. The interviews aimed at determining the key elements and processes of the investigated online firestorms, e.g., behaviors and decision-making functions of the individuals involved.

To specify the concrete properties and behaviors of the agents in our model, the inputs as well as the outputs, we used the data gained from our main case, the "customize-it" community run by the German toy manufacturer "Playmobil." We reviewed the transcripts of our interviews with the conflict participants and analyzed some 10.000 forum posts. In addition to these qualitative analyses, we conducted a network analysis in order to be able to validly reproduce the "customize-it" community's network structure. For example, the thresholds in the activation function of agents participating in positive/negative word of mouth were estimated by counting the past messaging behavior (see Table 1). The level of activity (as measured by the number of written messages) determined the assigned threshold.

The individual weights of the variables "perceived affectedness of oneself," "perceived affectedness of peers," and "perceived affectedness of the customer community" were specified using semantic analysis of the forum posts during the conflict. Two coders coded the data independently from each other and achieved a high interrater reliability. In order to assign the agents with the role of either "opinion leader" or "follower," we conducted a content analysis of the forum posts of every member of the "customize-it" community and validated the independent coders' evaluations with data from a network analysis on the agent's "betweenness centrality" (Table 2).

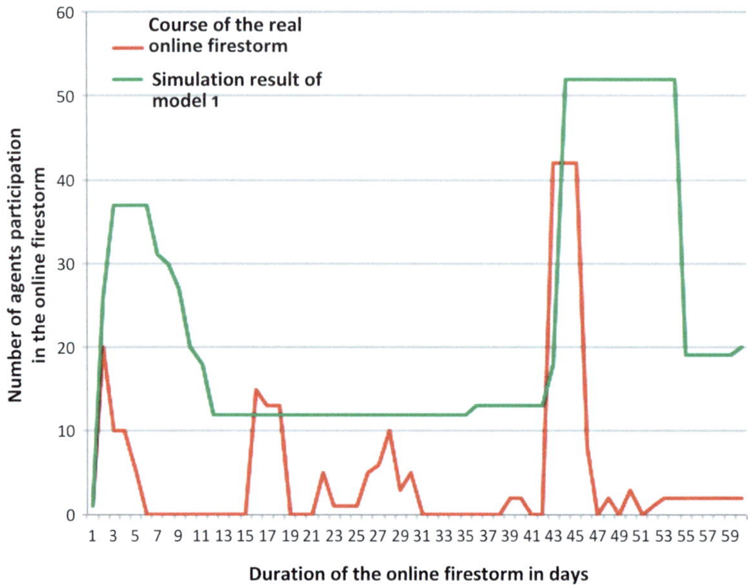

Fig. 1 Comparison of the real course of the "Playmobil" online firestorm with simulation results of model 1

We fed our model with the empirical data gathered by the procedures explained above. The simulation results of this very first model already rebuilt the online firestorm in the "customize-it" community very well (see Fig. 1). In the final step, we calibrated the input parameters to further improve the validity of our model. We iteratively adapted the agents' decision-making rules (except the weights and thresholds of the individuals, which had been derived empirically) and the opinion formation parameters (e.g., the impact of an "opinion leader") to reproduce the exact course of the real conflict [8].

3 Simulation Results

After validating the model, we manipulated the values of the input variables in order to analyze the functioning and effectiveness of firm-controlled supporters in pacifying online firestorms in different scenarios. Specifically, we ran sensitivity analyses by (1) varying the role of active firm-controlled supporters within the online customer community ("follower" vs. "opinion leader") and (2) varying the quantity of active firm-controlled supporters within the online customer community (zero, one, or ten). To check for the robustness of our simulation results, we also varied the characteristics of the online customer community. We simulated scenarios in which we varied (3) the initial average attitude toward the company within the online customer community and (4) the percentage of agents perceiving themselves, their peers, or the online customer community itself potentially affected by the company's misbehavior (20% of the agents perceiving an affectedness vs. 80% of the agents perceiving an affectedness). In total we modeled four communities that differ in their characteristics, which combined with the different agent scenarios sum up to 20 different, simulated conflicts (4*5). For a summary of all scenarios, see Table 3.

Our simulations of different conflicts aimed at investigating the effectiveness of the "infiltration" strategy in pacifying online firestorms. In particular, we looked if and how the properties of firm-controlled supporters (their number and their roles) as well as properties of the online customer community (average attitude toward the company prior to the conflict and the proportion of members affected by the misbehavior of the company) moderate the effect of an "infiltration" strategy on the intensity, speed, and length of online firestorms as well as on the average attitude toward the company after the end of such conflicts.

To illustrate the potential power of "infiltration" strategies, we compared two extreme scenarios with each other (see Fig. 2). Both scenarios display the diffusion of a negative word of mouth during an online firestorm within the "customize-it" community at three different points of time (t_0, outset of the conflict; t_1, middle of the conflict; t_2, end of the conflict). The black dots represent community members with a negative attitude toward the company; the gray dots represent neutral community members, and the white dots represent community members with a positive attitude toward the company (which in this case was "Playmobil"). The bigger dots are the complainers (black) and the firm-controlled supporters (white). In scenario A, 80% of all the community members are potentially affected by the misbehavior of the company (the prohibition of customization activities) and therefore have a higher likelihood to proactively participate in the online firestorm. No firm-controlled supporters are actively trying to pacify the online firestorm. Scenario B differs from scenario A only with respect to this last aspect. In this scenario, ten "opinion leaders" are actively and constantly spreading a positive word of mouth in order to reduce the negative consequences of the online firestorm and positively affect its dynamics.

Comparing the two scenarios shows the significant effect of an "infiltration" strategy. In scenario A (no firm-controlled supporters participating in the conflict), a large proportion of all community members hold a negative attitude toward the company at the end of the online firestorm. In scenario B (ten "opinion leaders" actively spreading a positive word of mouth), most community members hold a positive attitude toward the company at the end of the online firestorm. Furthermore, the number of community members with a negative attitude toward the company has even decreased as compared with the status at the outset of the conflict.

In order to more carefully analyze the effectiveness of the "infiltration" strategy, we compared the variance of all output variables of all scenarios where firm-controlled supporters were absent and all scenarios where they were present. We conducted analyses of variance in SPSS including post hoc tests. Table 3 provides a summary of the results of the ANOVAs.

Table 3 ANOVA results

DV: Final negatives (FN), negative peak (NP), maximum increase in negatives (MI), and negative man-days (NM)[a]

		Original case (80% affected)				Positive community				Negative community				20% affected			
# firm controlled supporters	n[b]	FN (SD)	NP (SD)	MI (SD)	NM (SD)	FN (SD)	NP (SD)	MI (SD)	NM (SD)	FN (SD)	NP (SD)	MI (SD)	NM (SD)	FN (SD)	NP (SD)	MI (SD)	NM (SD)
0	384	241.79	270.93	127.53	7790.16	96.94	161.55	75.03	3390.89	285.56	310.49	105.95	11081.80	49.81	146.72	45.13	3332.29
		(63.02)	(33.27)	(27.18)	(2737.12)	(55.56)	(46.45)	(29.63)	(1743.32)	(50.40)	(13.99)	(17.41)	(3442.45)	(42.15)	(71.31)	(31.60)	(2340.45)
1 follower	192	188.24	246.69	123.08	6989.90	60.70	138.06	66.91	2584.71	240.70	299.53	105.60	10177.42	42.12	135.93	43.75	2890.79
		(52.49)	(42.69)	(29.65)	(2911.19)	(37.44)	(38.19)	(24.77)	(1315.70)	(48.38)	(16.84)	(17.50)	(3427.49)	(39.49)	(66.84)	(30.26)	(2153.78)
1 opinion leader	192	151.50	232.10	119.76	6033.63	52.30	125.73	62.99	2200.62	218.88	292.86	104.74	9527.87	34.56	126.83	42.14	2528.82
		(59.16)	(44.98)	(29.62)	(2564.44)	(50.95)	(43.80)	(24.53)	(1393.16)	(56.60)	(20.06)	(18.64)	(3191.80)	(27.97)	(60.29)	(29.31)	(1728.48)
10 followers	192	69.80	184.76	98.09	3968.52	16.03	75.36	35.31	1034.51	133.51	254.57	96.32	6443.63	44.37	107.29	32.47	2438.44
		(22.67)	(40.59)	(33.74)	(2458.32)	(8.44)	(15.70)	(8.31)	(356.05)	(26.09)	(12.23)	(16.18)	(1600.62)	(33.21)	(55.20)	(22.93)	(1877.80)
10 opinion leaders	192	54.88	132.94	71.35	2804.32	29.90	65.80	32.56	1238.38	102.11	216.43	76.65	4844.27	21.30	66.26	21.94	1184.41
		(67.77)	(71.96)	(42.18)	(2899.11)	(58.17)	(54.74)	(26.74)	(1630.52)	(50.90)	(35.70)	(27.00)	(1969.39)	(10.10)	(19.20)	(14.47)	(653.79)
Effect f. —c supporters[c]		$p < .001$	$p < .001$	$p < .001$	$p < .001$	$p < .001$	$p < .001$	$p < .001$	$p < .001$	$p < .001$	$p < .001$	$p < .001$	$p < .001$	$p < .001$	$p < .001$	$p < .001$	$p < .001$
		$F = 767.13$	$F = 335.09$	$F = 104.35$	$F = 201.55$	$F = 338.45$	$F = 377.63$	$F = 213.00$	$F = 287.44$	$F = 706.58$	$F = 441.35$	$F = 56.37$	$F = 239.76$	$F = 41.96$	$F = 88.80$	$F = 32.35$	$F = 73.05$
Effect of role of f. —c supporters[d]		$p < .001$	$p < .001$	$p < .001$	$p < .001$	n.s.	$p < .01$	$p < .1$	n.s.	$p < .001$	$p < .001$	$p < .001$	$p < .001$	$p < .001$	$p < .001$	$p < .01$	$p < .001$
		$F = 22.24$	$F = 47.87$	$F = 27.93$	$F = 21.78$	$F = .66$	$F = 8.87$	$F = 2.89$	$F = .77$	$F = 25.43$	$F = 63.76$	$F = 38.92$	$F = 20.98$	$F = 49.63$	$F = 35.11$	$F = 10.22$	$F = 39.93$
Effect of # of f. —c supporters[e]		$p < .001$	$p < .001$	$p < .001$	$p < .001$	$p < .001$	$p < .001$	$p < .001$	$p < .001$	$p < .001$	$p < .001$	$p < .001$	$p < .001$	$p < .05$	$p < .001$	$p < .001$	$p < .001$
		$F = 732.67$	$F = 409.18$	$F = 205.02$	$F = 242.94$	$F = 114.14$	$F = 429.40$	$F = 368.28$	$F = 185.45$	$F = 1010.94$	$F = 990.65$	$F = 146.19$	$F = 456.57$	$F = 6.06$	$F = 123.49$	$F = 74.33$	$F = 49.97$
ANOVA with all groups		$p < .001$	$p < .001$	$p < .001$	$p < .001$	$p < .001$	$p < .001$	$p < .001$	$p < .001$	$p < .001$	$p < .001$	$p < .001$	$p < .001$	$p < .001$	$p < .001$	$p < .001$	$p < .001$
		$F = 498.36$	$F = 308.93$	$F = 112.25$	$F = 133.09$	$F = 118.74$	$F = 229.94$	$F = 141.55$	$F = 119.40$	$F = 620.17$	$F = 816.83$	$F = 86.94$	$F = 188.08$	$F = 17.27$	$F = 63.22$	$F = 28.07$	$F = 40.75$

[a]The results presented here are averages based on 100 repetitions of each model

[b]Number of scenarios

[c]Analysis of variance comparing all scenarios with firm-controlled supporters (group 1) with scenarios without firm-controlled supporters (group 2)

[d]Analysis of variance comparing all scenarios with firm-controlled opinion leaders (group 1) with all scenarios with firm-controlled followers (group 2)

[e]Analysis of variance comparing all scenarios with 1 firm-controlled supporter (group 1) with all scenarios with 10 firm-controlled supporters (group 2)

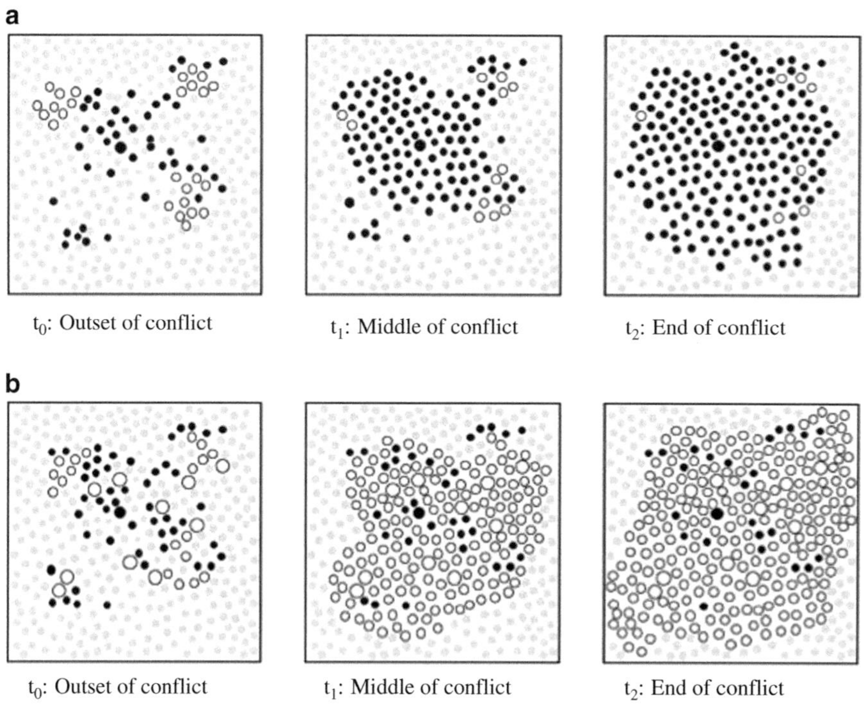

Fig. 2 Effectiveness of "infiltration" strategy with ten opinion leaders. (**a**) Scenario 1, no agent, original community, 80% affected. (**b**) Scenario 2, ten opinion leaders as agents, original community, 80% affected

Generally, our simulations show that "infiltration" strategies significantly help pacify online firestorms. More specifically, we found that if firm-controlled supporters proactively spread a positive word of mouth about the company in the course of an online firestorm, the number of community members with a negative attitude toward the company (final negatives – NM) at the end of the conflict is significantly lower than in scenarios without firm-controlled supporters participating in the conflict. This finding is not only true in the original "customize-it" community ($p < .001$; $F = 767.13$) but also in customer communities in which only 20% of all community members were affected by the customization prohibition ($p < .001$; $F = 41.96$), in customer communities with a positive average attitude toward the company ($p < .001$; $F = 338.45$), and in customer communities with a negative average attitude toward the company ($p < .001$; $F = 706.58$). Our findings further show that even a single firm-controlled supporter (follower) reduced the number of community members with a negative attitude toward the company by 23% on average (across the four different community settings). In the case of ten firm-controlled supporters with opinion leader status, the average reduction of community members with a negative firm attitude was as high as 67%.

"Infiltration" strategies do not only affect the public opinion within the online customer community but also the dynamics of the online firestorm. We find that firm-controlled supporters actively participating in the conflict reduce the intensity (in terms of the negative peak – NP) and the duration (in terms of negative man-days – NM) of the online firestorm. The effects are relatively strong in the original "customize-it" community with F-values of 335.09 (intensity) and 201.55 (duration) and highly significant (all p-values < .001). Again, these patterns hold true for all investigated settings of the online customer community (see Table 3). The potential reduction of the duration of an online firestorm is a particularly interesting finding. If only one follower acted as a firm-controlled supporter, the duration of the conflict could be reduced by 14% on average. Ten opinion leaders acting as firm-controlled supporters managed to reduce the duration of the online firestorm by an average of 62% (these numbers are again an average across all community settings). The effect of "infiltration" strategies on the third output variable describing the dynamics of online firestorms, the speed (in terms of the maximum increase in negatives), is not as clear though. The speed of the diffusion of negative word of mouth is much lower in scenarios in which firm-controlled supporters are active ($p < .001$, $F = 94.33$). However, post hoc tests revealed that this overall result is caused by only those scenarios in which ten firm-controlled supporters were active. In scenarios with only one firm-controlled supporter, the speed of the conflict did not significantly decrease. This finding is robust across all four variations of the "customize-it" community.

To investigate the importance of the role(s) of firm-controlled supporters, we compared the variances of those scenarios in which they were "opinion leaders" with those in which they were "followers." Our ANOVAs reveal that firm-controlled supporters with an "opinion leadership" position exert a much stronger influence on the average attitude toward the company within the community after the online firestorm than "followers" ($p < .001$, $F = 22.24$).

To investigate the importance of the number of firm-controlled supporters participating in the conflict, we compared the variances of those scenarios in which only one firm-controlled supporter was active with those in which ten firm-controlled supporters were active. The ANOVA results suggest that the quantity of firm-controlled supporters spreading a positive word of mouth about the company is positively associated with the average attitude toward the company within the online customer community at the end of the conflict ($p < .001$, $F = 732.67$). However, an interesting finding of our study is the fact that even a single firm-controlled "follower" has the capability of significantly affecting the community's average attitude toward the company after the conflict.

4 Conclusion

The purpose of this paper was to examine the effectiveness of employing firm-controlled supporters to pacify online firestorms via a case-based agent model. Our model suggests that employing firm-controlled supporters in the course of online firestorms positively affects the consequences and dynamics of an online firestorm. Interestingly, our simulations show that even a single firm-controlled supporter may significantly help pacify the conflict. This finding contradicts existing literature that points to the necessity to have a group of individuals employed in spreading a positive word of mouth in order to effectively influence opinions within social entities [5, 31]. Even more interesting and also contradicting existing literature, such a firm-controlled supporter does not even have to be an "opinion leader" within the community in order to calm a conflict within a community. This surprising insight can be explained by the specific behavior of firm-controlled individuals. They show "extremist" behavior, i.e., they continuously spread their (positive) opinion without getting influenced by other individuals themselves [4, 14, 17, 31]. Because of their persistence, they repeatedly expose a significant number of "neutral" individuals with their opinion, which increases their chance to exert influence over their peers. In other words, firm-controlled individuals from the periphery of a community make up for their decentralized network position and their lack of authoritativeness by a higher interaction frequency. This allows even followers to calm down online firestorms effectively [7, 10]. These insights are not only theoretically interesting but also of high relevance to the practical realm. They indicate that companies do not have to take a huge effort in order to infiltrate an online customer community by firm-controlled supporters. It is not necessary to "buy" the key members (i.e., the "opinion leaders") within such communities (which is not only expensive but also bears the risk of being caught trying to influence the community's opinion). Instead, even a single firm-employed supporter entering the online customer community may be able to significantly influence the course of the conflict or calm and deescalated its outcome. Clearly, these statements only hold true for the Playmobil customize-it case, and further case studies and general agent-based and classical models are necessary to broaden these assertions.

References

1. Amaral, L. A. N., Scala, A., Barthelemy, M., & Stanley, H. E. (2000). Classes of small-world networks. *Proceedings of the National Academy of Sciences, 97*(21), 11149–11152.
2. Ashforth, B. E., & Mael, F. (1989). Social identity theory and the organization. *Academy of Management Review, 14*(1), 20–39.
3. Boero, R., & Squazzoni, F. (2005). Does empirical embeddedness matter? Methodological issues on agent-based models for analytical social science. *Journal of Artificial Societies and Social Simulation, 8*(4).
4. Dahlander, L., & Wallin, M. W. (2006). A man on the inside: Unlocking communities as complementary assets. *Research Policy, 35*(8), 1243–1259.
5. Dellarocas, C. (2006). Strategic manipulation of internet opinion forums: Implications for consumer sand firms. *Management Science, 52*(10), 1577–1593.
6. Gebauer, J., Füller, J., & Pezzei, R. (2013). The dark and the bright side of co-creation: Triggers of member behavior in online innovation communities. *Journal of Business Research, 66*(9), 1516–1527.
7. Gilbert, N. (2008). *Agent-based models* (Vol. 153). Los Angeles: Sage.

8. Goldenberg, J., Libai, B., & Muller, E. (2010). The chilling effects of network externalities. *International Journal of Research in Marketing, 27*(1), 4–15.
9. Hafner-Burton, E. M., & Montgomery, A. H. (2006). Power positions international organizations, social networks, and conflict. *Journal of Conflict Resolution, 50*(1), 3–27.
10. Hegselmann, R., & Krause, U. (2002). Opinion dynamics and bounded confidence models, analysis, and simulation. *Journal of Artificial Societies and Social Simulation, 5*(3), 1–33.
11. Hennig-Thurau, T., Gwinner, K. P., Walsh, G., & Gremler, D. D. (2004). Electronic word-of-mouth via consumer-opinion platforms: What motivates consumers to articulate themselves on the internet? *Journal of Interactive Marketing, 18*(1), 38–52.
12. Hennig-Thurau, T., Malthouse, E. C., Friege, C., Gensler, S., Lobschat, L., Rangaswamy, A., & Skiera, B. (2010). The impact of new media on customer relationships. *Journal of Service Research, 13*(3), 311–330.
13. Hyde, C. K. (1986). Undercover and underground: Labor spies and mine management in the early twentieth century. *The Business History Review, 60*(01), 1–27.
14. Katz, E., & Lazarsfeld, P. F. (1955). *Personal influence*. New York: Free Press.
15. Levin, D. Z., & Cross, R. (2004). The strength of weak ties you can trust: The mediating role of trust in effective knowledge transfer. *Management Science, 50*(11), 1477–1490.
16. Miller, G. I. (1987). Observations on police undercover work. *Criminology, 25*(1), 27–46.
17. Miller, K. D., Fabian, F., & Lin, S.-J. (2009). Strategies for online communities. *Strategic Management Journal, 30*(3), 305–322.
18. North, M. J., & Macal, C. M. (2007). *Managing business complexity: Discovering strategic solutions with agent-based modeling and simulation*. Oxford: Oxford University Press.
19. Pace, S., Balboni, B., & Gistri, G. (2014). The effects of social media on brand attitude and WOM during a brand crisis: Evidences from the Barilla case. *Journal of Marketing Communications*, (ahead-of-print), 1–14.
20. Pfeffer, J., Zorbach, T., & Carley, K. M. (2013). Understanding online firestorms: Negative word-of-mouth dynamics in social media networks. *Journal of Marketing Communications, 20*(1–2), 117–128.
21. Rand, W., & Rust, R. T. (2011). Agent-based modeling in marketing: Guidelines for rigor. *International Journal of Research in Marketing, 28*(3), 181–193.
22. Rempel, M. W., & Fisher, R. J. (1997). Perceived threat, cohesion, and group problem solving in intergroup conflict. *International Journal of Conflict Management, 8*(3), 216–234.
23. Rouhana, N. N., & Fiske, S. T. (1995). Perception of power, threat, and conflict intensity in asymmetric intergroup conflict Arab and Jewish citizens of Israel. *Journal of Conflict Resolution, 39*(1), 49–81.
24. Sprague, R., & Wells, M. E. (2010). Regulating online buzz marketing: Untangling a web of deceit. *American Business Law Journal, 47*(3), 415–454.
25. Smith, T., Coyle, J. R., Lightfoot, E., & Scott, A. (2007). Reconsidering models of influence: The relationship between consumer social networks and word-of-mouth effectiveness. *Journal of Advertising Research, 47*, 387–397.
26. Stich, L., Golla, G., & Nanopoulos, A. (2014). Modelling the spread of negative word-of-mouth in online social networks. *Journal of Decision Systems*, (ahead-of-print), 1–19. doi:10.1080/12460125.2014.886494.
27. Turner, J. C. (1975). Social comparison and social identity: Some prospects for intergroup behaviour. *European Journal of Social Psychology, 5*(1), 1–34.
28. Van Noort, G., & Willemsen, L. M. (2012). Online damage control: The effects of proactive versus reactive webcare interventions in consumer-generated and brand-generated platforms. *Journal of Interactive Marketing, 26*(3), 131–140.
29. Ward, J. C., & Ostrom, A. L. (2006). Complaining to the masses: The role of protest framing in customer-created complaint web sites. *Journal of Consumer Research, 33*(2), 220–230.
30. Wasko, M. M., & Faraj, S. (2005). Why should I share? Examining social capital and knowledge contribution in electronic networks of practice. *MIS Quarterly*, 35–57.
31. Watts, D. J., & Dodds, P. S. (2007). Influentials, networks, and public opinion formation. *Journal of Consumer Research, 34*(4), 441–458.

Soft Systems Agent-Based Methodology: Multi-methods Approach Between Soft Systems Methodology and Agent-Based Modeling

Santi Novani and Lidia Mayangsari

Abstract The world is now moving toward a pluralistic approach of combining several methods, both hard and soft, within an intervention multimethodology. It currently shows an upward movement for social researcher to employ methodologies grounded on environmental analysis and ecosystem modeler, known as hard system, or preferably on participatory processes, stated as soft system. Both approaches surely have their own positive and negative points to reckon with the complexity of systems from various perspectives. However, a major existing gap is the lack of a clear methodology for integrating these two approaches. This paper synthesizes a promising integration in soft systems methodology (SSM) and agent-based modeling (ABM) that hereinafter referred to as soft systems agent-based methodology (SSABM). SSABM emerges from the limitations of soft systems dynamic methodology (SSDM), which combines SSM and system dynamics (SD) approach. The theoretical and methodological assumption in SSM, ABM, and SD is briefly discussed, and critiques for each method are presented. Regarding their theoretical and methodological approach, this paper shows a comparison between SSABM and SSDM in dealing with decision-making.

Keywords Soft systems methodology • Agent-based modeling

1 Introduction

The plethora of varied methods during the 1970s and 1980s has boosted the possibility of combining tools in intervention which is well known as multimethodology [12]. Following that idea, this paper has a notion that some of the strengths and weaknesses of two widely used system-based approaches promise a prominent future collaborating together. The integration result considered as a new approach linking both approaches like a mutual symbiosis. Furthermore, this paper has been inspired by its predecessor known as SSDM which pronounces SSM with SD as its

S. Novani (✉) • L. Mayangsari
School of Business and Management, Institut Teknologi Bandung, Bandung, Indonesia
e-mail: snovani@sbm-itb.ac.id; lidia.mayangsari@sbm-itb.ac.id

U.S. Putro et al. (eds.), *Agent-Based Approaches in Economics and Social Complex Systems IX*, Agent-Based Social Systems 15, DOI 10.1007/978-981-10-3662-0_13

crucial partner [11]. SSDM can be essentially employed to demonstrate that much can be gained in a systemic intervention [15].

However, the problems in social world have now been evolved to the higher level of complexity. SD has placed a strong highlight on the notion of counterintuitive behavior of complex dynamic systems. That happens due to the intricate interplay between many interrelated factors and the nonlinearity of their relationships [1]. Although SD simulation modeling and analysis are essential for robust policy design [8], the dynamic behavior of complex systems becomes somewhat practically difficult to predict because of SD static structure [4].

People are also a main feature in complex dynamic world that SD just slightly pays attention to. This paper introduces ABM to be paired with SSM in a more comprehensive way due to its theoretical and methodological approach. ABM itself is a simulation methodology that employs an implicit worldview [18] that is very close to a network of independent actor point of view. Complex ABM dynamics models the behavior of the world through agent's employment [7]. SSM and ABM have similarity in terms of both systemic approaches based on human activity systems and not reluctant to cultural political aspects. In this area of complex dynamics, those two similarities are insight to combine SSM and ABM to be the key of complex adaptive systems in the future. This paper claims the combination of those two widely used system-based approaches as soft systems agent-based modeling (SSABM).

To demonstrate this claim, this paper convinces the difference between SSDM and SSABM with their epistemological and ontological assumptions underpinning their correspondent paradigm. To be well presented, this paper is structured as follows. In Sect. 2, the SSABM methodology is explained from its epistemological, ontological, and methodological principles to be compared with SSM, ABM, and SSDM. Then, the synthesizing role of SSABM and stage differentiation with SSDM is stated in Sect. 3. Furthermore, Sect. 4 concludes the learning points of this paper and further researches proposed.

2 Defining SSABM

2.1 SSM Paradigm

Soft systems methodology (SSM) is a process-oriented approach for channeling debate about situations characterized by messy ill-structured problems with multiple perspectives [5]. Soft systems thinking may be distinguished from "hard" in that SSM locates the use of systems concepts in the process of inquiry rather than "objective" observations of the real world. SSM has been used in a great variety of contexts and with a great variety in styles of intervention. Effective use of SSM requires a situation where participants have sufficient time to share and learn from each other and a high enough level of trust to discuss their preferences and requirements openly.

Use of SSM begins with the decision that a problem situation exists that could benefit from a complete rethinking. Checkland defines a problem as a condition characterized by a state of mismatch that eludes precise definition and may be expressed simply as a state of unease. Participants are recruited to bring together as diverse a group of stakeholders as possible to consider the situation in a systemic way.

SSM is done through a seven-stage process. The stages do not need to be taken in any particular order. Those seven stages are (1) looking at the unstructured problem, (2) structuring a problem statement, (3) formulating root definitions, (4) building conceptual models, (5) comparing models and "reality," (6) defining changes, and (7) taking action.

2.2 *ABM Paradigm*

Agent-based modeling (ABM) is now currently used in social science to describe and understand the dynamics of social and economic systems (cite). This method emerged from political science case [2] and then becomes generally applicable to study another complex system such as biological systems [20], ecologies [10], and spatial systems [17].

ABM is well suited for social science objective since it stands on the following two foundations: (1) the system is composed of interacting agents, and (2) the system exhibits emergent properties, that is, properties arising from the interactions of the agents driven by a set of rules which govern agents' behavior [3]. The interaction between the agents exposes complex system behavior whose emergent dynamic properties cannot be explained by analyzing its component parts. When the interaction of the agents is contingent on past experience, and especially when the agents continually adapt to that experience, mathematical analysis is typically very limited in its ability to derive the dynamic consequences. In this case, ABM might be the only practical method of analysis. Depart from these properties, ABM serves flexibility and capacity to represent social phenomena by means of intuition, objects, and relations, where basic mathematical formulation or classical micro-simulation tends to fail in explaining the phenomena.

Agent-based model can be considered as bottom-up approach that relies on interacting agent to build system behavior or properties. It consists of a space, framework, or environment in which interactions take place and a number of agents whose behavior in this space is defined by a basic set of rules and by characteristic parameters. Models can be spatially explicit, i.e., agents are associated with a specific location from which they may or may not be able to move. Not all models need to be spatially explicit if the location does not play a role.

Framework that is composed of states and relations becomes the main component of the ABM. When a system is in operation, the states of its agents and their relations with one another are changing through time. These collective changes in state can be considered a process (also called the systematic behavior). Similarly, a particular agent and its changes in state can also be considered treated as a process or behavior. Moreover, this kind of process may interact with the any relevant external component and alter the environment.

2.3 SSABM and SSDM Paradigm in Contrast

Both SSABM and SSDM have the same master platform as seen in their typical stages adopted from SSM. The most significant disparity from both of them is the utilization of SD for SSDM and ABM for SSABM.

SD methodology focuses on the structure composed by the interactions of the elements, flows and levels, between them (Scholl). SD methodology process mainly follows three steps [11]. The first step is understanding situation or problem definition which emphasizes on the need to ensure of the real issue or behavior of a system. Next is model conceptualization or model building which means to structurize the system and constructs computer simulation. The third final step is to test different policies scenarios and what if situation. However, SD is an aggregate and macro-level model of a system. It works in top-down approach way with abstract state variables and equations, so one of SD limitations is practically difficult to reach micro-parts of the system. SD models have a fixed dynamic structure. Consequently, it is not able to reproduce the process of moving from one structure to another. Another limitation is SD models typically include "soft" variables that are difficult to translate into numerical values.

ABM aims to look at global consequences of individual or local interactions in a given space [16]. It is a method for studying systems exhibiting two properties: the system is composed of interacting agents, and the system exhibits emergent properties. Emergent properties are properties arising from the interactions of the agents that cannot be deduced simply by aggregating the properties of the agents [3]. The term agent has several definitions. Agent is an individual with a set of characteristics or attributes. Agent is also a set of rules governing behaviors or decision-making capability and protocols for communication. Agent also responds to the environment and interacts with other agents in the system. ABM talks about bottom-up approach that involves agents and their rules of interaction. One standout point from ABM is that agents are diverse and heterogeneous and that makes ABM interesting.

The disparities brought by SD and ABM principles set SSDM apart from SSABM in their ontology, epistemology, and also methodology as written in Table 1.

Table 1 Principle of SSM, ABM, SSDM, and SSABM

	SSM	ABM	SSDM	SSABM
Ontological principle	Systems are not assumed to exist in the real world; social world of attributing meaning	Systems exist in the real world; emphasizes meanings based on the changing interconnections among autonomous, heterogeneous social agents	Systems are not assumed to exist in the real world. The social world has meaning for observer	Systems are not assumed to exist in the real world. The social world interaction and behavior have meaning for observer
Epistemological principle	Interpretivist, phenomenological and (possible) hermeneutical claims	Postpositivist, model-centered including logic algorithm and simple probabilities	Interpretivist, rationalistic, phenomenological, and hermeneutical assumptions	Interpretivist, rationalistic, sociopsychological, phenomenological, and hermeneutical assumptions. Bounded rational and KISS logic algorithm included
	Describes the real world in epistemological terms (verbs)	Describes the world in terms of emerging properties of behavior and interactions (verbs, noun, and adverbs)	Describes the real world in epistemological and ontological terms (verbs, nouns, and adverb)	Describes the real world ontologically from the individual interactions and epistemologically from the model built by the governed behavior (verbs, nouns, and adverb)
	Separation of the real world and systems thinking world; systematicity is in the process	Separation of the real and system world is not very clear	Separation of the real world and systems thinking world is clear; divides SSDM systems thinking world into two: (1)problem situation-oriented systems thinking world and (2)solving problem situation-oriented systems thinking world	Separation of the real world and systems thinking world is clear; (1)problem situation-oriented systems thinking world and (2)solving problem situation-oriented systems thinking world
Methodological stages	Systemic approach based on “logical” linked human activity systems	Systemic approach based on interrelations among individuals with controlled behavior in a given area and time	Systemic approach based on “logical” linked human activity systems and “rational” cause-effect relationship	Systemic approach based on “logical” linked human activity systems and interrelations among individuals with controlled behavior in a given area and time

(continued)

Table 1 (continued)

	SSM	ABM	SSDM	SSABM
	Seeks for cultural feasible and systematically desirable changes in the real world	Gains insight from cultural feasible as an input for agent behavior rules	Looks for cultural feasible and systematically desirable changes in the real world	Looks for cultural feasible and systematically desirable changes in the real world
	It is a problem and solving oriented methodology	It is a problem-solving-oriented approach	It is a problem and solving oriented methodology	It is a problem and solving oriented methodology
	Unable to measure and assess the possible changes by itself through the time	Able to foresee, measure, and assess the changes by itself through controlled behavior in a period of time	Able to measure and assess the problematic and improved situation by itself through time	Able to foresee, measure, and assess the changes by itself through controlled behavior in a period of time
	Clearly establishes the "what" and "how" transformation process performed or to be performed in the real world, to "improve" it	The "what" and "how" transformation process implemented in the real world is clear	Clearly establishes two transformation processes: (1) one which explains "what" is the problem situation and "how" it behaves and (2) the other which explains "what" and "how" should be the transformation process to "improve" or "alleviate" that problem and "how" the improved situation should behave	Clearly establishes two transformation processes: (1) one which explains "what" is the problem situation and "how" it behaves and (2) the other which explains "what" and "how" should be the transformation process to "improve" or "alleviate" that problem and "how" the improved situation should behave
	It is not a dialectic approach	It is not a dialectic approach	It is a dialectic approach	It is a dialectical approach
	It finishes with a learning process from the application of the whole methodology in an informal way	It finishes with a learning process of the model building in an informal way	It finishes with a formal process of learning from three positions: (a) from the problematic view of the problematic sit (SSDM's World 2), (b) from the solucionatic view of the problematic sit (SSDM's World 3), (c) from the implementation process in the real world (SSDM's World 1)	It finishes with a formal process of learning from three positions: (a) from the problematic view of the problematic sit (SSABM's World 2), (b) from the solucionatic view of the problematic sit (SSABM's World 3), (c) from the implementation process in the real world (SSABM's World 1)

3 Stages in SSABM

SSABM stages are basically adopted from SSDM stages which have been fully developed by Rodriguez-Ulloa and Paucar-Caceres. There are ten stages in SSDM methodology involved in three different worlds: the real world (red boxes in Fig. 1), the problem situation-oriented systems thinking world (blue boxes in Fig. 1), and the solving situation-oriented systems thinking world (light blue boxes in Fig. 1). The first world will briefly describe the whole situation in the reality. After getting to know what the reality is, and then the second world is a term for problem situation appreciation and holistic behavior recognition. In the third world, which also happens to be the dialectical side, we study about how systems thinking can "solve," "finish," or "alleviate" the problem situation and propose it in the solving-situation world. In SSABM methodology, the ten stages also involve three worlds to work on. However, there are differences in stages 4 and 7 that focus on model building; see Fig. 1.

3.1 Stage 1: Unstructured Problem Situation

This stage was borrowed from soft systems methodology. In SSM the researchers begin with a real-world problem. They study the situation in a highly unstructured

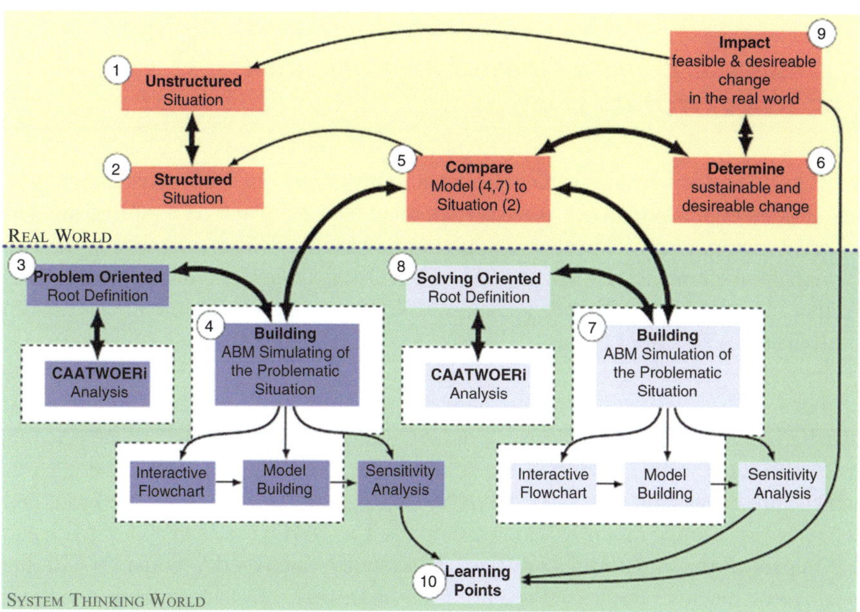

Fig. 1 Step in soft systems agent-based methodology

way. Then, they develop some models represent that situation. An unstructured problem is any situation where a mismatch can be perceived between two people or more. The purpose of this stage is to acknowledge, explore, and define the situation in some way. According to Checkland, this stage doesn't define the problem but assesses the general area that interests us. We can choose to open out the boundary of the situation to sweep in more aspects of the situation. It will not be possible to define the problem or its setting with any precision, and, in any event, the different people involved will have different ideas.

3.2 Stage 2: Structured Situation

This stage also comes from SSM. The situation needs to be expressed to make researchers understand the problem easily. Checkland suggests using a rich picture. It can help to organize complex situations and identify underlying problems by including all relevant elements and stakeholders of a system. The principle behind rich pictures is that drawing detailed representations of problematic situations can be a tool to organize thoughts, better understand what is going on in the system, and identify root problems. This holistic thinking may help to see relationships and connections that may otherwise be missed [19]. Any technique may be used in drawing a rich picture, and they may be used in any situation. But structure, process, and concerns are elements that should be included to build a rich picture.

3.3 Stage 3: Problem-Oriented Root Definitions (CAATWOERI Analysis)

CATWOE analysis which is clients, agents, transformation, Weltanschauung, owners, and emergent properties is already defined by Checkland as a part of the soft systems methodology. However, combining SSM and ABM means adding the two fundamental elements of ABM to SSM which are attributes and rules of interactions so that CATWOE analysis becomes CAATWOERI analysis. Below is the table to describe CAATWOERI (Table 2).

In SSM, we call actors as the people involved in the implementation of the changes in the system. But agents here are a feature of agent-based modeling. It can be people, companies, projects, assets, animals, products, and so on. Agents have attributes, often described by simple rules, and interact with other agents, which influence their behaviors. CAATWOERI analysis also needs the input and the nature of change inputs to become outputs. The worldview is the big picture and the impact of the transformed system. The system is analyzed to come up with the positive or negative impact representing the real world. In agent-based modeling, we need to define how our model can cover the problem. There are the decision-

Table 2 CAATWOERI definitions

Clients	The victims or beneficiaries of T
Agents	Those who do T
Attributes	Information owned by each agent, for example, characteristic of an object (person, thing, etc.)
Transformation	Input à output
Weltanschauung	The worldview that makes the T meaningful in context
Owners	Those with the power to stop T
Emergent properties	Overall behavior of the system emerges from the interactions
Rule of interaction	Decision-making algorithms or behavioral probabilities

makers who have the authority to make the changes, stop the program, set the goal of simulation of the model, and so on. By modeling agents individually, the full effects of the diversity of their attributes and behaviors can be observed to emerge the behavior of the whole system. Since agent-based modeling usually runs in computer simulation, the researcher needs to make an algorithm as a rule interaction among agents involved in the system.

3.4 Stage 4: Agent-Based Model Building in the "Problematic Situation"

In this stage, researchers build a model based on the real situation. As we know, an agent model has three elements, a set of agents (attributes and behaviors), a set of agent relationships, and the agent environment, which is agents that interact with their environment in addition to other agents. A model developer must identify, model, and program these elements to create an agent-based model without any intervention. A model should be a specific problem that interests the researcher to be solved. The agents, the decision-makers, and the entities involved should be cleared as well. Actually, there are many approaches to designing and implementing agent-based models. For example, [14] discuss both design methodologies and selected implementation environments in depth. For bottom-up methodology like agent-based models, highly iterative design methodologies seem to be most suitable and effective for practical development.

3.5 Stage 5: Comparison Between Stages 4 and 2

After the model is built, the researcher must compare to the real situation. Modeling is the process of building an abstraction of a system for a specific purpose. If an abstraction does not represent its modeling target, it would be inappropriate to call it model. So, the researcher should make sure the model is represented the reality. All the things are not directly modeled, but an abstraction of the target should be modeled [6]. In essence, what distinguishes one modeling paradigm from another is precisely the way we construct that abstraction from the observed system. According to [9], a model may contain errors and artifacts. An error is a mismatch between what the modeler thinks the model is and what it actually is. In contrast to errors, artifacts relate to situation where there is no mismatch between the modeler thinks a model is and what it actually is. The mismatch here is between the set of assumptions in the model that the modeler thinks are producing a certain phenomenon and the assumptions that are the actual cause of such phenomenon.

3.6 Stage 6: Culturally Feasible and Systematically Desirable Changes Determination

This stage backs to SSM paradigm. In the ideal situation, the changes should cover all aspects of the system being analyzed and the viewpoints of all agents in the model. However, projects in the real world are always subject to schedule and resource constraints. As a result, researchers have to make decisions to prioritize the various requirements. In contrast to hard systems approaches that tend to emphasize that change should be systematically feasible and culturally desirable, in this stage, change should be systematically desirable and then culturally desirable.

3.7 Stage 7: Agent-Based Model Building in the "Solving Situation"

In this stage, researchers build a model to solve situation. A model should be a specific problem that interests the researcher to be solved. Based on the model from the real situation that already developed in stage 4, and also made some desirable change from stage 5 and 6, we make an improvement model in order to solve the situation.

3.8 Stage 8: Solving Situation-Oriented Root Definitions (CAATWOERI Analysis)

CAATWOERI analysis also needs the input and the nature of change inputs to become outputs. The worldview is the big picture and the impact of the transformed system. The system is analyzed to come up with the positive or negative impact representing the real world. In agent-based modeling, we need to define how our model can cover the problem. There are the decision-makers who have the authority to make the changes, stop the program, set the goal of simulation of the model, and so on. By modeling agents individually, the full effects of the diversity of their attributes and behaviors can be observed to emerge the behavior of the whole system. Since agent-based modeling usually runs in computer simulation, the researcher needs to make an algorithm as a rule interaction among agents involved in the system.

3.9 Stage 9: Feasible and Desirable Changes Implementation in the Real World

To implement the "feasible and desirable" developments and/or changes in the real word, we need to put them in action. By changing the structure of the procedure and attitude, it can be easy to specify and easy to implement even though in the real word it might be difficult.

3.10 Stage 10: Learning Points

The learning point is the final and essential stage of the SSABM process as it reflects the whole stages involved in the process of solving problem of the situation. It accumulates the wrong and the right emerged from the attributes implanted in the agents along the process. Finally, it weights to which direction the decision for solution should go. The learning points provide many lessons to learn from the situation brought up in the process.

4 Conclusion and Further Research

This paper convinces the difference between SSDM and SSABM with their epistemological and ontological assumptions underpinning their correspondent paradigm. This research will also be furtherly equipped with service system science perspective as a new scientific discipline. Service science emerges as an interdisciplinary discipline, which combines all discipline of technological, engineering, and social system in order to achieve value cocreation [13]. As a real-world problematic situation, batik industrial cluster in Solo is chosen since it has unique characteristic which is composed of interaction between entities of service system, i.e., entrepreneurs, customers, government, and other supporting agencies. SSABM works in this field of application since SSM illustrates human activity model and ABM simulation sees the micro-interaction among the service system entities in batik cluster in Solo. Data are collected through interviews conducted with service science entities in Solo batik industrial cluster.

References

1. Akkermans, H. A. (2001). Renga: A systems approach to facilitating inter-organisational network development. *System Dynamics Review, 17*(3), 179–194.
2. Axelrod, R. J. ConQict Res. 41 (1997) 203 (reprinted in R. Axelrod, *The complexity of cooperation*, Princeton University Press, Princeton, 1997).
3. Axelrod, R. (2006). Advancing the art of simulation in the social sciences. In J.-P. Rennard (Ed.), *Handbook of research on nature inspired computing for economics and management*. IGR: Hershey.
4. Borschev, A. and Filippov, A. (2004). From system dynamics and discrete event to practical agent based modelling: Reasons, techniques, tools. In *22nd international conference of the system dynamics society*. Oxford, UK. online: www.systemdynamics.org/conferences/2004/SDS_2004/PAPERS/381BORSH.pdf
5. Checkland, P. B. (1999). *Systems thinking, systems practice*. Chichester: Wiley.
6. Edmond, B. (2001). The use of models-making MABS actually work. In S. Moss & P. Davidsson (Eds.), *Multi-agent-based simulation, lecture notes in artificial intelligence 1979* (pp. 15–32). Berlin: Springer-Verlag.
7. Fonseca, M. G. D., & Zeidan, R. M. (2004). Epistemological considerations on agent-based models in evolutionary consumer choice theory. *Emergence: Complexity and Organization, 6*(3), 4–8.
8. Forrester, J. W. (1961). *Industrial dynamics*. Cambridge, MA: MIT Press.
9. Galan, J. M., Izquierdo, L. R., Izquierdo, S. S., Santos, J. I., del Olmo, R., Lopez-Peredes, A., & Edmonds, B. (2009). Errors and artefacts in agent-based modeling. *Journal of Artificial Societies and Social Simulation, 12*(1), 1. http://jasss.soc.surrey.ac.uk/12/1/1.html.
10. Grimm, V., Revilla, E., Berger, U., Jeltsch, F., Mooij, W. M., Railsback, S. F., Thulke, H.-H., Weiner, J., Wiegand, T., & DeAngelis, D. L. (2005). Pattern-oriented modeling of agent-based complex systems: Lessons from ecology. *Science, 310*, 987–991.
11. Lane, D. C., & Olivia, R. (1998). The greater whole: Towards a synthesis of system dynamics and soft systems methodology. *European Journal of Operational Research, 107*(1), 214–235.
12. Munro, I., & Mingers, J. (2002). The use of multimethodology in practice – Results of a survey of practitioners. *The Journal of the Operational Research Society, 53*(4), 369–378.
13. Ng, I. C. L., Maull, R. S., & Smith, L. (2010). *Embedding the new discipline of service science*. University of Exeter Business School Discussion papers in Management, paper number 09/01 ISSN 1472–2939.
14. North, M. J., & Macal, C. M. (2007). *Managing business complexity: Discovering strategic solutions with agent-based modeling and simulation*. Oxford: Oxford University Press.
15. Paucar-Caceres, A., & Rodriguez-Ulloa, R. (2007). An application of soft systems dynamics methodology (SSDM). *Journal of the Operational Research Society, 58*(6), 701–713.
16. Reynolds, C. W. (1999). *Steering behaviors for autonomous characters*. Paper presented at the Game Developers Conference.
17. Sanders, L. (Ed.). (2007). *Models in spatial analysis, Geographical Information Systems series*. London: ISTE.
18. Shannon, R. E. (1975). *Systems simulation the art and science*. Englewood Cliffs: Prentice-Hall.
19. Wageningen, U. R. (Ed.). (2008). *Multi-stakeholder processes. Ghana, 2008*. Wageningen: Wageningen University and Research Centre.
20. Walker, B. H., Holling, C. S., Carpenter, S. R., & Kinzig, A. (2004). Resilience, adaptability and transformability in social–ecological systems. *Ecology and Society, 9*(2), 5. [online] URL: http://www.ecologyandsociety.org/vol9/iss2/art5
21. Zhang, T. I. (2010). Agent-based systems design for virtual organisations formation. In: *ICAART2010: Proceedings of the 2nd international conference on agents and artificial intelligence,* Valencia, January 22–24, 2010.

An Analysis of the Characteristics of a Proactive Exploration Behavior Model Based on Three-Dimensional Virtual Space Experiments

Toshiyuki Kaneda, Kohei Okamoto, Tatsuto Suzuki, and Masaki Tamada

Abstract EVA is an exploration behavior model and claims to simulate "natural movement" in Space Syntax Theory. However, for example, in the proactive exploration behaviors in the depa-chika, the customers select foods while freely circulating. This paper focuses on examination in a Lessened Strokes Exploration Model (hereinafter, LSEM), developed by the authors as a proactive exploration behavior model incorporating the lessened strokes principle and imperfect recall by the removal of "footprints." Through gaming experiments to confirm characteristics of LSEM, we demonstrate the potential of LSEM interpretations and extended version of the behavior models. Using a three-dimensional virtual space system, LSEM was compared with the analyzing data of gaming experiments and examining aspects of characterized human behavior. Concerning the gaming experiment in a 3D virtual space, LSEM shows a similar tendency.

Keywords 3D virtual space • Depa-chika • Proactive exploration behavior • Space syntax theory • LSEM • Gaming simulation

T. Kaneda (✉) • T. Suzuki
Nagoya Institute of Technology, Nagoya, Aichi, Japan
e-mail: kaneda@nitech.ac.jp; ttt.3916@gmail.com

K. Okamoto
ObayashiCo Ltd, Minato, Tokyo, Japan
e-mail: okamoto.kohei@obayashi.co.jp

M. Tamada
Kozo-Keikaku Engineering Co Ltd, Nakano, Tokyo, Japan
e-mail: tamada@kke.co.jp

U.S. Putro et al. (eds.), *Agent-Based Approaches in Economics and Social Complex Systems IX*, Agent-Based Social Systems 15, DOI 10.1007/978-981-10-3662-0_14

1 Research Background and Objectives

EVA (Exosomatic Visual Architecture) developed by UCL is a well-known agent model representing free exploration behaviors such as strolling and claims to simulate the "natural movement" described in Space Syntax Theory [1–4].

In contrast to these behaviors, behaviors to explore something like an object or target can be described as proactive exploration behaviors and classified into two types: "way-finding exploration behavior" and "article-finding exploration behavior." Let us consider the typical article-finding exploration behavior observed in today's Japanese depa-chika, a department store basement with its crowds of shoppers milling around the food products section: some are looking for specific items, and others are totally undecided and browsing. Amidst this flurry of activity, for reasons of efficiency, shoppers have a tendency to avoid a pathway that they have already taken, but at the same time, shoppers do duplicate their routes due to their imperfect mental map or their imperfect register and recall. However, there are very few discussions among researchers concerning models representing such article-finding exploration behaviors.

This research examined our newly devised Lessened Strokes Exploration Model (hereinafter, LSEM) [5, 6] developed as a proactive exploration behavior model incorporating the lessened strokes principle (behavior with as few strokes as possible) and a lapse of memory function (imperfect human memory) and firstly conducted a simulation experiment to confirm its characteristics. Next, using a three-dimensional (3D) virtual space system, an experiment with participants was carried out to analyze the behavior rules found in human proactive exploration behaviors. Moreover, LSEM was compared with the data collected from the gaming experiment to analyze and examine which aspects of human behavior characteristics are represented by LSEM.

2 Proactive Exploration Behaviors and Their Model

2.1 *What Is LSEM?*

Our LSEM examined in this research retains the memory of walk history or the so-called footprints on a walking plane, so that agents, as much as possible, avoid following in their own footprints. In other words, it is a pedestrian model in which agents behave so as to minimize "the number of strokes." Moreover, the disappearance of these footprints enables the operational handling of a "lapse of memory." Figure 1 shows the experiment space and boundaries. In accordance with the behavior algorithm shown in Fig. 2, the LSEM agents enact behaviors until they satisfy the allocated number of products to be purchased.

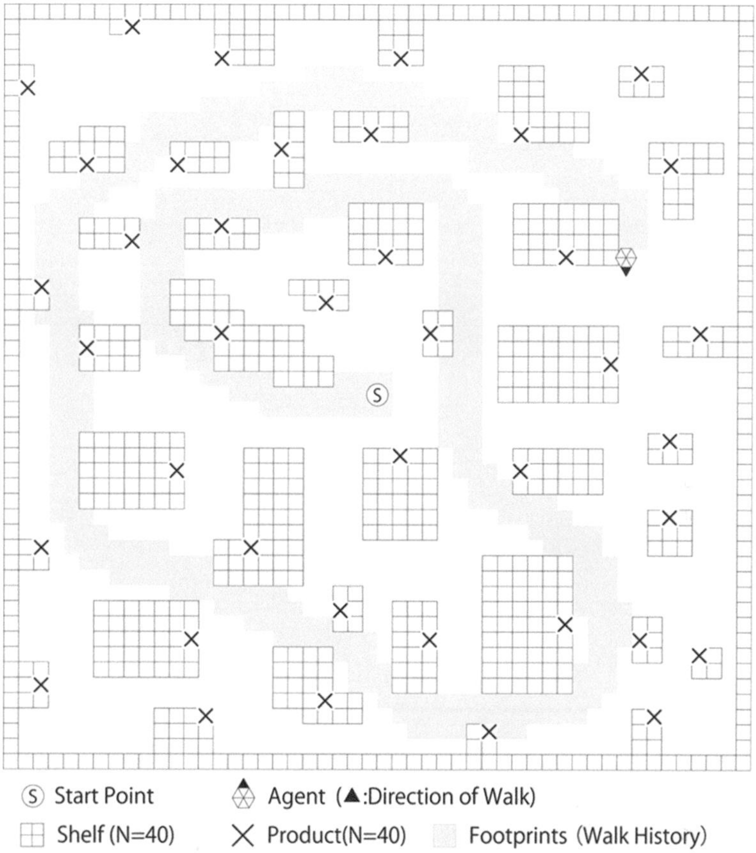

Fig. 1 Experiment space for proactive exploration behaviors

2.2 Characteristics of LSEM

This section explains in detail the behavior principle of LSEM. Firstly, let us consider an agent located on a space represented by fine grids and facing in the direction of walking. The whole of the agent's field of vision centering on the walking direction is divided into a fan of equal angles; each of the fan sectors is described as a "bin" (division unit sector). Each bin represents a candidate direction to be considered when changing the walking direction. The agent compares the sight lines of several candidate bins, selects the deepest line (direct distance to a shelf or wall), moves forward in that direction one unit length per one time step, and continues to repeat this rule. Here, a set of sight line candidates is equal to an isovist, which is a region that is made up of a visual step depth (VSD) value of 1 in the visibility graph analysis of Space Syntax Theory. It is possible to consider

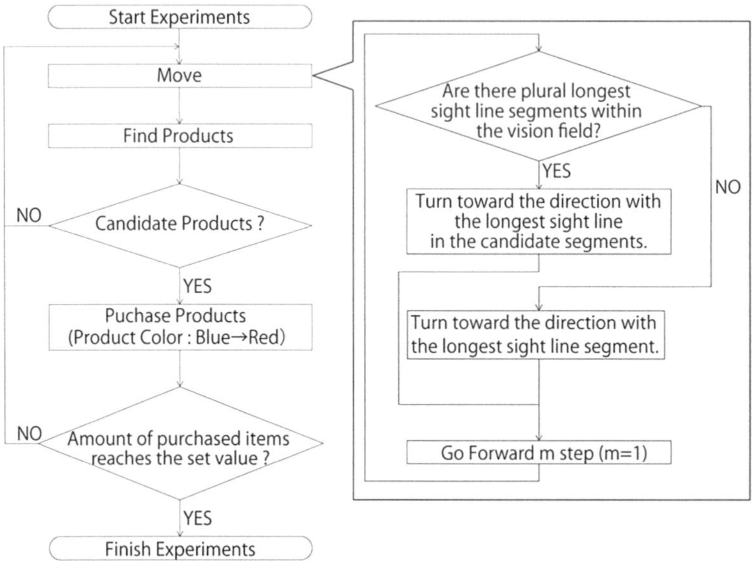

Fig. 2 Behavior algorithm of LSEM (Lessened Strokes Exploration Model)

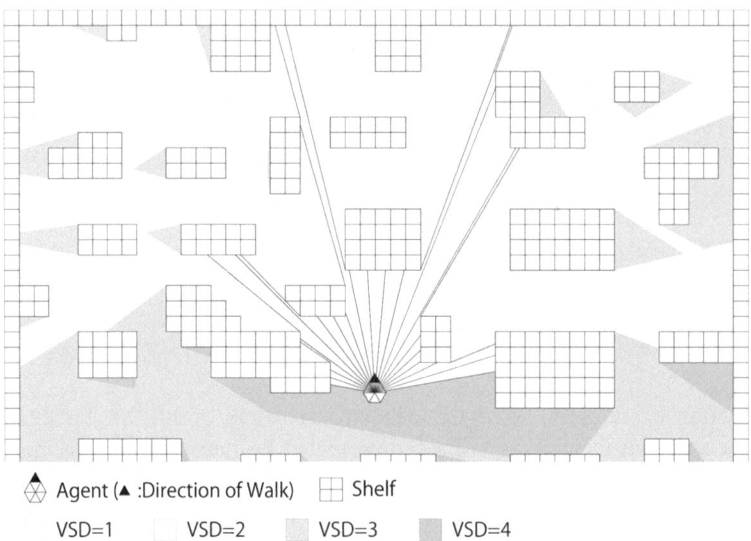

Fig. 3 Explanation of the agent's field of vision and visual step depth

extension to a case where the visual step depth value is 2 or 3; in this regard, the sight line length can be replaced with the sum total of lengths (Fig. 3).

In the LSEM, every time an agent goes forward one unit in length, they register a "footprint" on the walking plane. At each decision phase of the walking direction,

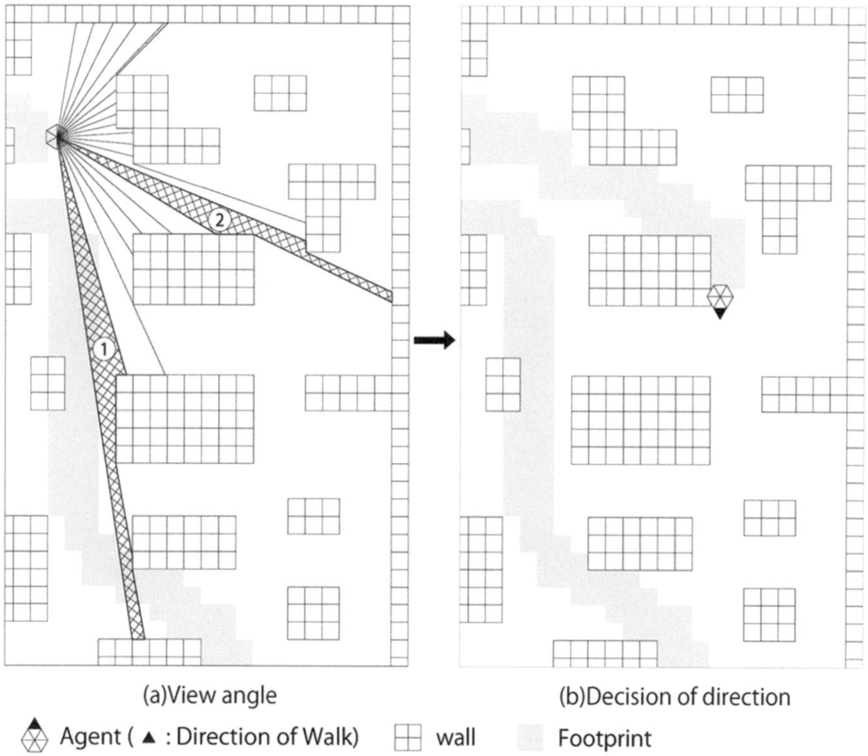

Fig. 4 Selection of a walking direction under footprint constraints

in cases of the sight line blocked by "footprint," the sight line must be replaced by one that is a partial line to the block. It is called the sight line segment. If this line is not blocked by any footprints, the sight line segment coincides to the sight line. Then, turn toward the direction with the longest segment in the vision field. If there exist plural candidate segments, their sight line is used as the tie-break principle (Fig. 4). With regard to the "footprints," as a token recording that the agent has passed through a sufficiently narrow path, a row of three tiles (the middle tile behind the direction of walking and the adjacent tile to the right and left) are registered on the walking plane.

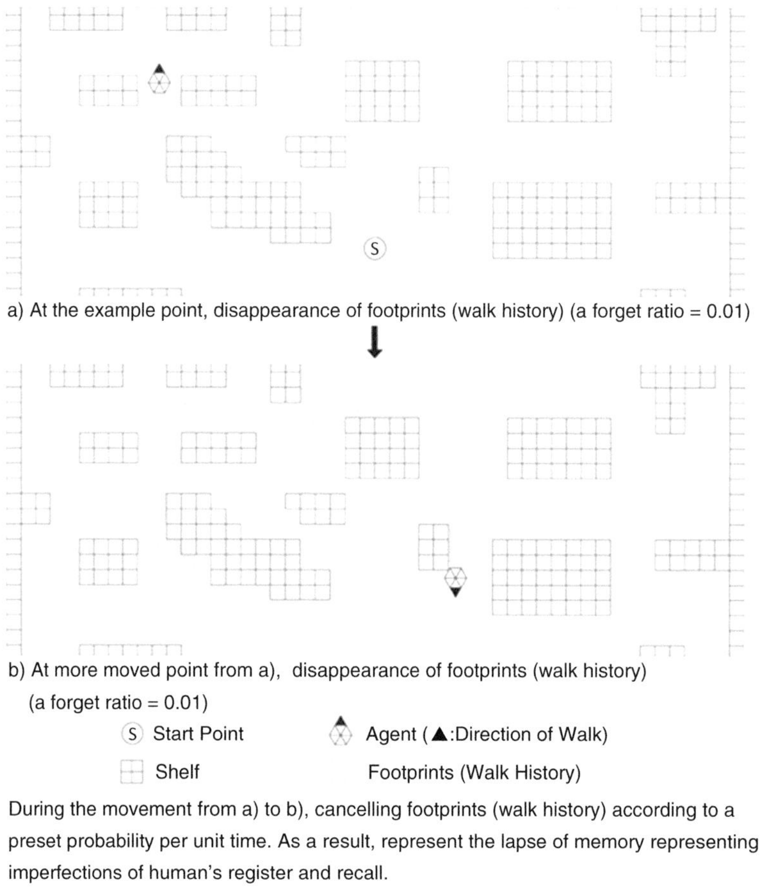

Fig. 5 Diagrammatic explanation of a lapse of memory by the disappearance of footprints (walk history)

2.3 LSEM with Lapses of Memory

A "lapse of memory" was introduced into LSEM, by cancelling footprints (walk history) recorded on an agent's transit route according to a preset probability per unit time, and this study examines the characteristics of the model by conducting simulation analysis (Fig. 5).

To examine exploration behaviors in an unfamiliar space, a set of sight line candidates is a region that is made up of a visual step depth (VSD) value of 1 in the visibility graph analysis of Space Syntax Theory, and simulations were conducted for five cases with a forget ratio of 0, 0.01, 0.02, 0.03, or 0.05; Figure 6 shows the results. This simulation terminates when an agent has approached products at 30 of the 40 product points shown by crosses in Fig. 1, and in all cases, simulations

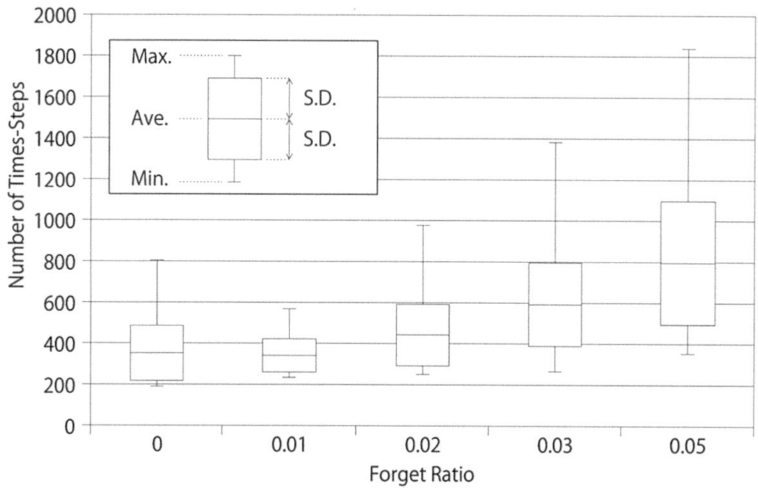

Fig. 6 Forget ratios and walking behavior redundancy

were conducted 50 times. When the number of walking steps is focused on as a redundancy indicator, there is a clear tendency toward a monotonic increase in relation to the forget ratio. In addition, in LSEM, in the case with a sufficiently high forget ratio (footprints are not recorded), such a model can be considered as including not only proactive exploration behaviors but also free exploration behaviors. By comparing with the gaming experiment data, basically it has now become possible to determine which VSD case in LSEM is best suited to the relevant behavior.

3 Gaming Experiment in 3D Virtual Space

3.1 Creation and Characteristics of a 3D Virtual Space

To compare the data on agent walking behavior in the LSEM simulations with the data on behaviors in 3D virtual space, a virtual space was created using two types of 3D design tool software: StoneyDesigner and SketchUp. For comparison, the same number of shelves and products and the same layout as in LSEM were applied (Fig. 1).

This paragraph explains the experimental system. On the experiment screen, 2D and 3D maps are displayed, but participants can only see the 3D map. Each participant controls an on-screen pedestrian (corresponding to an agent in LSEM) in the virtual space by using four keyboard keys: E (move forward), S (rotate left), D (rotate right), and X (move back). Footprints are recorded on the 2D map in the same manner as in LSEM. To conduct analysis, concurrently with the end of the

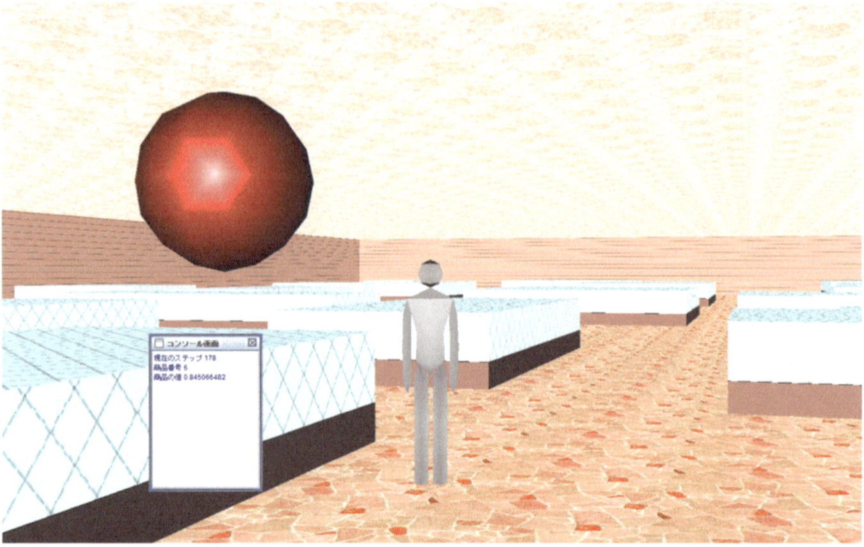

Fig. 7 A screen example when a product is found in the 3D virtual space experiment

experiment, the walking record, footprint record, and CSV file (number of time steps, coordinates for each number of time steps, and number of duplicate tiles) are saved to subsequently calculate the number of walking steps.

3.2 Outline of the 3D Virtual Space Gaming Experiment

With 25 student participants, the experiment was conducted using the situation shown in Fig. 1. When a pedestrian moves to any point within the product recognition zone, a red sphere appears along with the product price, and each time, the participant assesses whether the product is a candidate or not (Fig. 7). When the participant has found three out of five candidate products, the experiment ends.

4 Examination by Comparing the 3D Virtual Space Gaming Experiment and LSEM Simulation Results

4.1 Examination of the Forget Ratio Observed in the 3D Virtual Space Gaming Experiment

This section verifies whether the LSEM simulation with a forget ratio of zero (0) (hereinafter, forget ratio 0 case) shows a tendency close to the gaming experiment

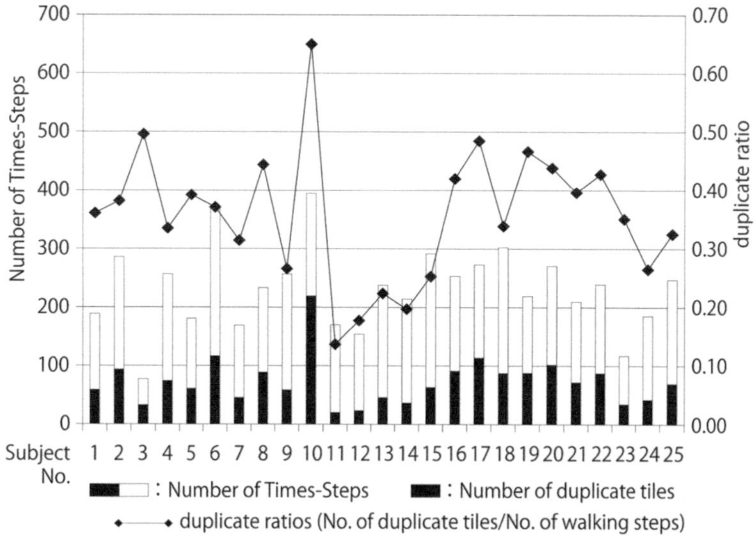

Fig. 8 Behavior experiment duplicate ratios: no. of duplicate tiles/no. of walking steps

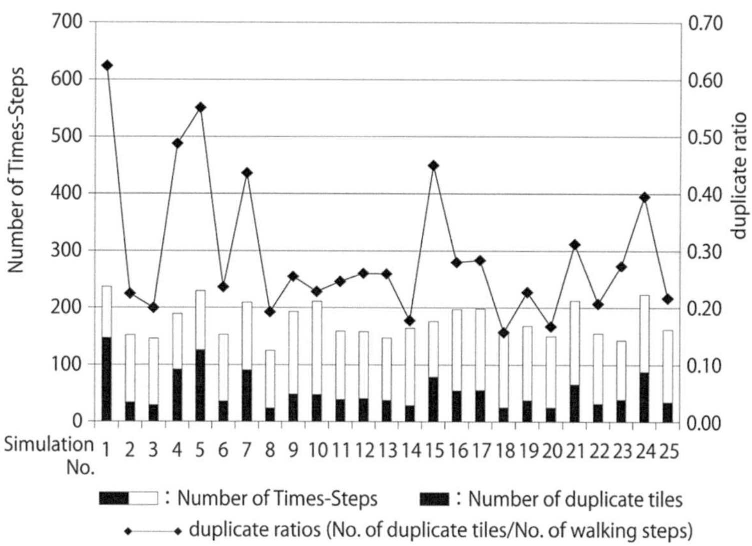

Fig. 9 LSEM simulation (forget ratio 0 case) duplicate ratios: no. of duplicate tiles/no. of walking steps

data. Since the gaming experiment was conducted under a condition of no mental map, the VSD for the forget ratio 0 case was set as 1. The measurement results of the gaming experiment are given in Fig. 8 and those of the LSEM forget ratio 0 case in Fig. 9. First, the statistical test had not found the difference of their time steps within

Table 1 Verification of the difference in the duplicate ratio average values

	Ave. (S.D.)	Test result
The 3D Virtual space subject experiment ($N = 25$)	0.305	$p = 0.712$ (No significant difference was found at a significant level of 10%)
	(0.096)	
Simulation of LESM (ratio of forget : 0 case, $N = 25$)	0.293	
	(0.122)	

a significant level of 10%. Then, the duplicate ratio was calculated by dividing the number of duplicate tiles by the number of walking steps as an indicator to show how often the same route was followed, and a verification was carried out for the difference in the average values between the gaming experiment and the forget ratio 0 case; Table 1 shows the results. The average value for the gaming experiment was 0.305 and for the forget ratio 0 case was 0.293, with standard deviations of 0.096 and 0.122, respectively; the difference between them was reduced, and no significant difference was found at a significant level of 10%. These results allow us to consider that the forget ratio 0 case was close to the tendency of the data obtained by the gaming experiment.

4.2 *Verification of Applicability of the Footprint Constraints*

To verify the effectiveness of footprint constraints introduced to LSEM, using the data collected from the gaming experiment, coordinates were plotted every ten walking steps to extract behavior rules when selecting a walking direction. Figure 10 shows an example of classification for extracting behavior rules. The selection of the walking direction at each plot point was classified into the following three items: (1) deepest line direction with footprint constraints, which is the direction of the deepest line based on the footprint constraints; (2) deepest line direction without footprint constraints, which is the direction of the deepest line where the footprint constraints are ignored and the line crosses the footprints; and (3) others, which are neither (1) nor (2). Figure 11 shows the procedure to extract behavior rules. The results of the classified totals are shown in Table 2 and as a pie chart in Fig. 12. From among a total of 482 observation points, the deepest line direction with footprint constraints rule was applied the most (343 points (71.16%)), followed by others (102 points (21.16%)) and deepest line direction without footprint constraints (37 points (7.68%)). When the average number of direction selection times per person was examined, out of a total of 19.28, the deepest line direction with footprint constraints was 13.72, which greatly surpassed the deepest line direction without footprint constraints at 1.48 and others at 4.08. At the plot points exceeding 70%, the deepest line direction with footprint constraints was selected; this result could support the effectiveness of the footprint constraints in LSEM.

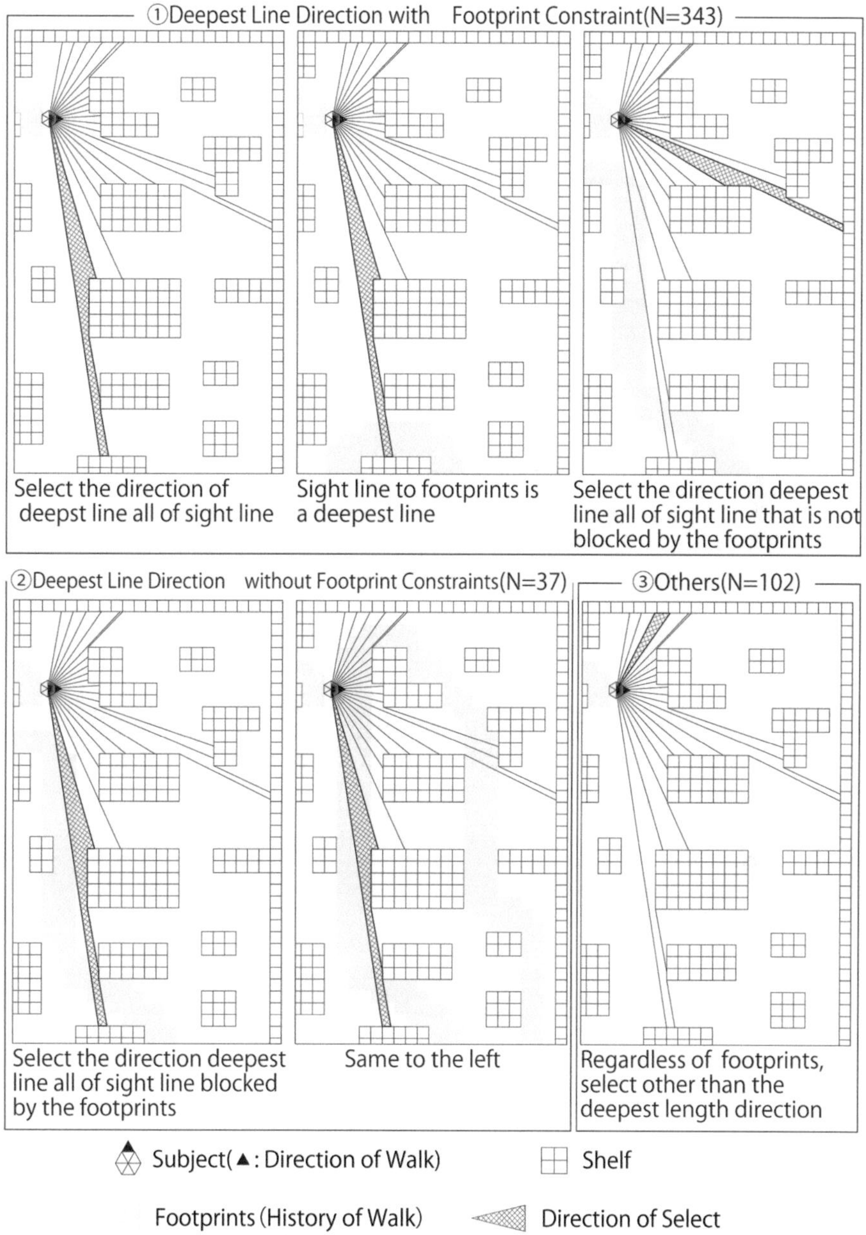

Fig. 10 An example of classification to extract behavior rules when selecting a walking direction

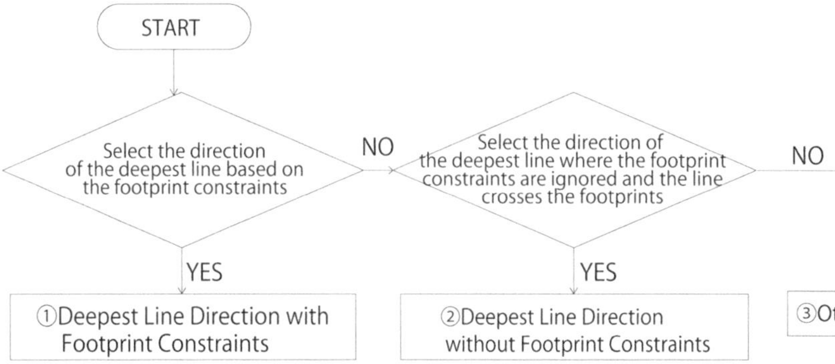

Fig. 11 Extraction of behavior rules when selecting a walking direction in the 3D virtual space experiment

Table 2 Number of times each behavior rule was selected every ten walking steps

	Number of times the selection of the appropriate direction	The average number of times per person
①Deepest Line Direction with Footprint Constraints	343	13.72
②Deepest Line Direction without Footprint Constraints	37	1.48
③Others	102	4.08
Total	482	19.28

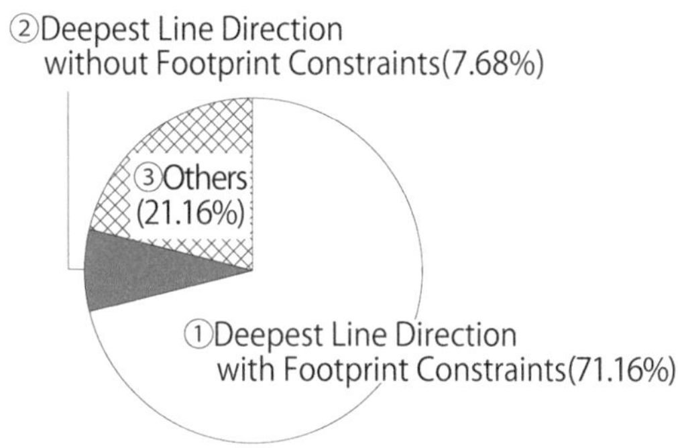

Fig. 12 Ratios of behavior rules in the walking direction's selection

5 Conclusion

The research first confirmed the characteristics of the proactive exploration behavior model LSEM, which attempts "lessened strokes" by introducing "footprint constraints," and then a comparison was made with the data from a gaming experiment in a 3D virtual space to verify which aspect of human behaviors is represented by LSEM. The obtained knowledge was as follows:

1) The simulations in LSEMver.2, to which a "lapse of memory" was introduced by removing footprints (walk history), clarified that the number of walking steps, a redundancy indicator, monotonously increased in relation to the forget ratio. Moreover, when the forget ratio is sufficiently high, it can be said LSEMver.2 is a model capable of representing free exploration behaviors.

2) Concerning the gaming experiment in a virtual space, LSEM (forget ratio 0 case) appears to show a similar tendency. In the gaming experiment, at 70% or more of the points, the deepest line direction with footprint constraints was selected; this suggests the effectiveness of a footprint constraint behavior rule in LSEM.

This paper confirmed that LSEM has basic characteristics as a proactive exploration behavior model representing "article-finding exploration behavior," which is different from free exploration behaviors. A setup for much more realistic sight angle and movement actions remains further problems.

Acknowledgment This work was supported by JSPS KAKENHI Grant Number 26282083 and 25240048.

References

1. Hillier, B., & Hanson, J. (1984). *The social logic of space*. Cambridge, MA: Cambridge University Press.
2. Penn, A., & Turner, A. (2001). Space syntax based agent simulation. In M. Schreckenberg & S. Sharma (Eds.), *Pedestrian and evacuation dynamics 2001* (pp. 99–114). Heidelberg: Springer-Verlag.
3. Turner, A., & Penn, A. (2002). Encoding natural movement as an agent-based system: An investigation into human pedestrian behavior in the built environment. *Environment and Planning. B, Planning & Design, 29*, 473–490.
4. Penn, A., & Turner, A. (2004). *A system and method for intelligent modelling of public spaces*, Patent Application GB0020850.4. UK Patent Office.
5. Kaneda, T. (2014). *A proactive exploration behavior simulation system for autonomous agents, international patent classification G06Q 10/00*. Japan Patent Office.
6. Kaneda, T., Okamoto, K., & Tamada, M. (2014). *Modeling and simulation of proactive exploration behaviors with imperfect recall, pedestrian and evacuation dynamics* (pp. 567–575). Delft: Elsevier.

Distributed Classifier System for Smart Home's Machine Learning

Mhd Irvan and Takao Terano

Abstract Recently, smart homes have become the center of Internet of Things (IoT) development. More and more connected devices are now parts of our house. These devices are considered as "smart" devices because they may be able to communicate with each other or with the Internet. However, they are not "smart" enough in a sense of learning. Most of the time, users have to manually interact with the devices to match their living style. This paper proposes a distributed machine learning method to automate the learning process for these devices to avoid making users repeat the manual interaction over and over again. Each device is able to self-learn to detect users' favorite settings during certain periods and then share this information among other self-learning smart devices to find patterns that may lead to the best combination of their settings that suit the living style of each member of the house. It implements a distributed classifier system that resulted in an algorithm that is small enough to run on lightweight single-board computers.

Keywords Machine learning • Smart home • IoT

1 Introduction

We hear many things about smart homes recently. Manufacturers are racing to build connected electronic devices for residential houses, essentially making them "smart" [1]. The term "smart" in this situation typically means that the devices are connected to the Internet and are able to communicate with each other [2]. However, they may not be smart in a sense of learning capability [3]. Most of the time, users have to manually interact with the devices using remote controls or their smartphones. They spend a lot of time adjusting and readjusting the settings again and again to match their preference.

Through this paper, we proposed a distributed machine learning method to automate the learning process for smart electronic devices. Each device is able to

M. Irvan (✉) • T. Terano
Department of Computational Intelligence and Systems Science, Interdisciplinary Graduate School of Science and Engineering, Tokyo Institute of Technology, Tokyo, Japan
e-mail: irvan@trn.dis.titech.ac.jp; terano@dis.titech.ac.jp

U.S. Putro et al. (eds.), *Agent-Based Approaches in Economics and Social Complex Systems IX*, Agent-Based Social Systems 15, DOI 10.1007/978-981-10-3662-0_15

self-learn to detect users' favorite settings during certain periods and then shares this information among other self-learning smart devices to find patterns that may lead to personalization of their settings that suit the living style of each member of the house, in essence helping users to avoid repeating the manual interaction with the devices frequently.

Our proposed method implements a distributed classifier system by combining a learning classifier system (LCS) method with a multi-agent system approach. The agents are tasked to solve subproblems assigned to them, rather than to solve the whole problems. By having multiple intelligent agents share knowledge among them, the overall learning process can be split into multiple smaller learning systems. This provides a machine learning method that is small enough to be run on lightweight single-board computers, typically used by smart home devices.

2 Proposed Method

Our proposed method implements a modified machine learning algorithm called learning classifier system (LCS) [4]. LCS has been so far recognized as a good approach for clustering [5] and multi-agent systems [6, 7]. LCS is essentially an intelligent agent having interaction with an environment. The agent first detects inputs from the environment. It then applies genetic algorithms (GA) [8] to search for a decision. Afterward, it takes an action based on the decision and finally gathers reinforcement learning (RL) [9] reward returned by the environment, basically informing the agent the quality of the action done by the agent (Fig. 1).

This paper is using a custom LCS called XOLCS [6, 7]. It is a multi-agent system (MAS) implementation of LCS. XOLCS follows the learning method of XCS [4], an LCS algorithm that puts more emphasis on accuracy compared to other LCS algorithms. XOLCS also takes inspiration from organizational learning [10] concepts to deal with the knowledge management within agents' classifier system, allowing them to retain good knowledge and transfer learning information among

Fig. 1 Illustration of a learning classifier system

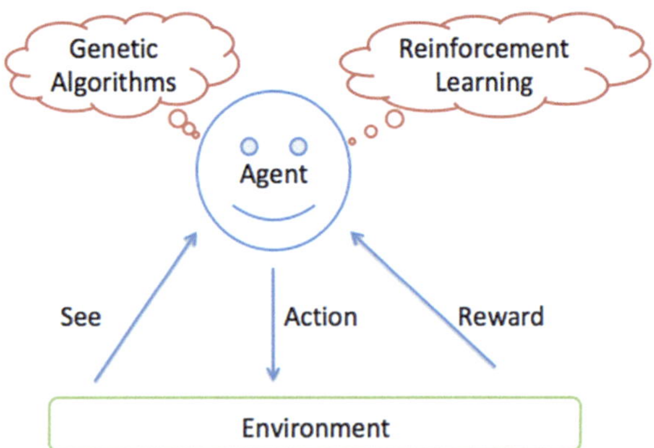

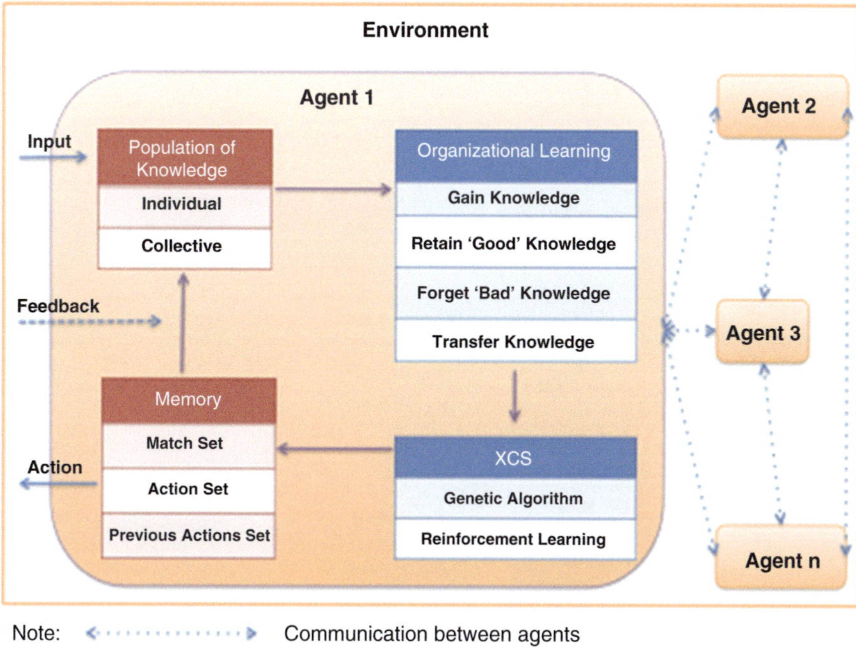

Fig. 2 XOLCS architecture

them. In XOLCS, each agent has its own classifier system. A classifier system is a learning process embedded within the pool of classifier (knowledge) rules that learns to distinguish good knowledge from bad knowledge through trial-and-error experience. Experience is gained from attempting action proposed by rules evolved by GA and receiving reward from RL (Fig. 2).

The learning process starts from detecting input fed by the environment. The agent then compares the input against the classifiers within its knowledge population, putting all rules that match the input into a match set. It then selects an action and puts all rules proposing that particular action into an action set. Finally, it executes the action and receives rewards from the environment. OL algorithms may be applied to modify the rule population before selecting an action, and GA operations can be applied to evolve the rules from the action set to give birth to new rules that can be inserted into the population (Fig. 3).

XOLCS makes its decision by calculating the prediction array $P(A)$ for the classifier C within match set $[M]$ using prediction ($C.p$) and fitness ($C.f$) values and then selects action with the highest prediction array's value:

$$P(A) = \frac{\sum_{C.a=a \wedge C \in [M]} C.p \times C.f}{\sum_{C.a=a \wedge C \in [M]} C.f} \tag{1}$$

$$A_{\max} = \arg\max_A P(A) \tag{2}$$

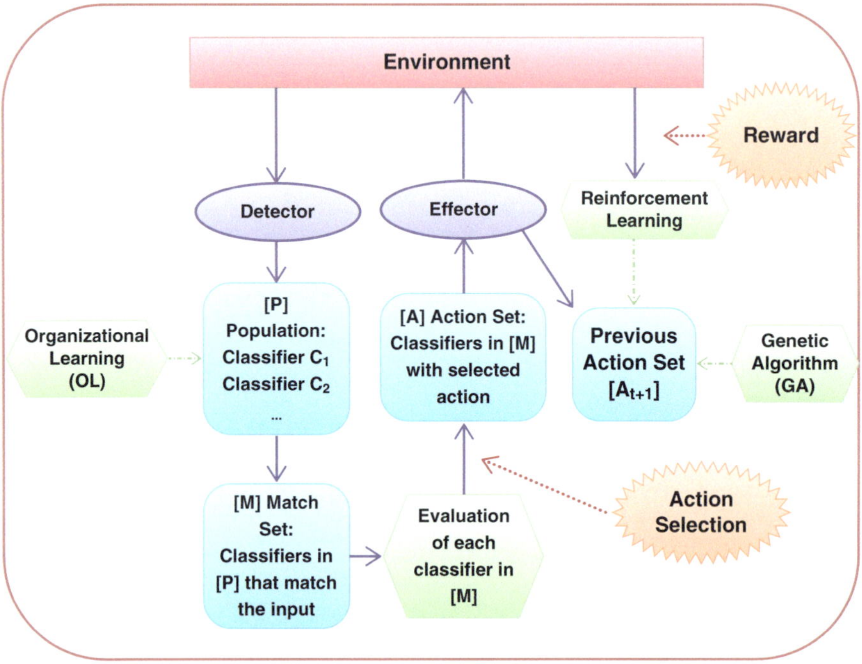

Fig. 3 XOLCS learning process

GA operations are applied when the average time for classifiers within the previous action set since the last occurrence of *GA* is higher than *GA* frequency parameter θ:

$$GA\_TRIGGER\,([A])\ \text{if}: \frac{\sum_{C\in[A]} t - C.g}{[A]\ \ n} > \theta \tag{3}$$

Similarly, *OL* operations are applied when the last occurrence of *OL* is greater than *OL* frequency parameter ϕ:

$$OL\_TRIGGER\,([P])\ \text{if}: \frac{\sum_{C\in[P]} t - C\ h}{[P]\ \ n} > \varphi \tag{4}$$

Lastly, reward R is given to the classifiers in the action set by deriving the maximum prediction within the prediction array $P(A)$, discounted by parameter γ:

$$Q = \max_A P(A) \tag{5}$$

$$R = r + \gamma Q \tag{6}$$

As mentioned before, knowledge is represented as a population of classifiers. In LCS, classifiers are defined as strings of condition-action rule, where the condition part specifies the situation when the classifiers are applicable and the action part specifies the action proposed by the classifiers. Additionally, each classifier also maintains prediction errors and accuracy properties to help track its performance overtime throughout the learning process.

The inputs represent various values that can possibly be detected by various sensors on the smart home devices. Although in practice there are many inputs that can be collected through various devices, for this early research presented on this paper, we limit the classifiers to the following inputs:

- Day: Represents the day the device is used
- Time: Represents the time when the device is used
- Room temperature: Represents the room temperature when the device is used
- Room brightness: Represents the lamp's brightness when the device is used
- TV volume: Represents the television's volume level when the device is used

In our simulation, each agent represents a device, having its own population set of classifiers. A classifier is represented as a string of inputs mentioned above and a setting prediction, acting as the chromosome for *GA*. The *GA* will then apply crossover and mutation algorithms to these classifiers to generate new classifiers, and RL's Q-learning algorithm is used to give feedback in a form of reward to the LCS regarding the quality of the classifiers. This reward is used to adjust the fitness value of each classifier and will affect the next decision made by LCS.

The results achieved by these learning agents can be used to smartly adjust the device settings without users' intervention. Users may change their preferences by adjusting the device manually, and the devices will relearn their new preferred settings.

3 Early Experiment

Because currently no public datasets that could represent the input types mentioned previously are available, and this research is still in its early stage, data is still being gathered. Thus, for this short paper, we conducted the experiment and analyzed our proposed method using a randomly generated dataset. These random datasets were generated through a Gaussian distribution with standard deviation from 0.01 to 0.05. The data was split across five different folds into two sets: (1) training set (80%) for XOLCS to learn and (2) validation set (20%) for XOLCS to predict.

Table 1 summarizes the precision and recall results for each fold, earned after ten rounds of simulation (averaged). For comparison, we also listed the result achieved by naïve Bayes classifier on our random dataset.

Our experiment result shows that LCS can be effective at predicting the values of the validation set after training XOLCS agents using the training set. Our method

Table 1 Experiment result

	Precision		Recall	
	XOLCS	Naïve Bayes classifier	XOLCS	Naïve Bayes classifier
First fold	75.8%	63.7%	79.7%	67.2%
Second fold	74.3%	62.2%	78.5%	66.1%
Third fold	73.6%	61.3%	77.9%	65.7%
Fourth fold	75.1%	63.4%	79.3%	67.6%
Fifth fold	74.7%	62.1%	78.2%	66.2%

was able to achieve better both precision and recall on each fold compared to standard naïve Bayes classifier method. On average, XOLCS produced more than 10% better precision and recall values. The combination of genetic algorithms and reinforcement learning managed to evolved classifiers to have better accuracy, and XOLCS correctly maintained the optimal set of classifiers throughout the simulation. Since our proposed method can be used as a general machine learning algorithm, the input types and the number of inputs can be flexibly changed. This can be convenient when future smart home devices come with additional settings, providing new data.

4 Conclusion and Future Work

Building a smart home is a big challenge. Many devices have only small computing power embedded in them with minimum capability to learn from complex data. By splitting the learning process between different devices, it is possible for one device to make use of data gathered from other devices. The multi-agent approach proposed in this paper allows such learning process to be executed in a distributed manner.

This paper presents a new research that is still in its early stage of work. The early experiment detailed above shows that XOLCS can be a feasible approach to coordinate machine learning across multiple smart home devices. Although it shows promising result, more testing on real data, algorithm optimization, and comparison to other methods is necessary.

References

1. Ricquebourg, V., et al. (2006). The smart home concept: Our immediate future. In *1st IEEE international conference on E-learning in industrial electronics* (pp. 23–28).
2. Harper, R. (2006). *Inside the smart home*, ISBN 1852338547, 9781852338541. London: Springer.
3. Rashidi, P., & Cook, D. J. (2009). Keeping the resident in the loop: Adapting the smart home to the user. *IEEE Transactions on Systems, Man and Cybernetics, Part A: Systems and Humans, 39*(5), 949–959.

4. Holland, J. H., et al. (2000). What is a learning classifier system? In *Learning classifier systems: From foundation to applications, lecture notes in computer science* (Vol. 1813, pp. 3–32). Berlin/Heidelberg: Springer.
5. Tamee, K., Bull, L., & Pinngern, O. (2006). A learning classifier system approach to clustering. In *Sixth International Conference on Intelligent Systems Design and Applications* (Vol. 1, pp. 621–626). IEEE.
6. Irvan, M., Yamada, T., & Terano, T. (2013a). Influence of organizational learning for multi-agent simulation based on an adaptive classifier system. *IEEJ Transactions on Electronics, Information and Systems, 133*(9), 1752–1761.
7. Irvan, M., Yamada, T., & Terano, T. (2013b). An on-the-fly learning framework for a fluctuating number of agents in a multi-agent system. In *Agent-based approaches in economic and social complex systems VII, Agent-based social systems* (Vol. 10, pp. 227–242). Tokyo: Springer.
8. Goldberg, D. E., & Holland, J. H. (1988). Genetic algorithms and machine learning. *Machine Learning, 3*(2–3), 95–99.
9. Kaelbling, L. P., Littman, M. L., & Moore, A. W. (1996). Reinforcement learning: A survey. *Journal of Artificial Intelligence Research, 4*, 237–285.
10. March, J. G. (1991). Exploration and exploitation in organizational learning. *Organization Science, 2*(1), 71–87.

Delivering Value of Composting and Recycling in Household Waste Management System: An Agent-Based Modeling Approach

Noorhan Firdaus Pambudi and Akbar Adhiutama

Abstract Community and government have different perspectives and roles in conducting household waste management system. On one hand, the government generates a policy to decrease waste to be sent to final disposal; on the other hand, the community will accept that, and without community involvement, sometimes the policy will not run well. Household waste management system policies focus on delivering value of composting and recycling in community. One of the policies is waste bank which lets the community turn their waste into money in the form of savings like in a bank. Efforts from different household waste supply chain parties such as government and communities must be coordinated. This research aims to find the exact coordination between government and communities in order to adopt household waste policies especially waste bank. This research considers possibilities to produce policies by determining community needs previously (bottom-up technologies). Agent-based modeling presents three kinds of community: the careless community (not willing to adopt waste management technologies/policies even it is profitable for them), arguing community (not willing to adopt waste management technologies/policies when it is not profitable for them), and adapting community (willing to adopt waste management technologies/policies whether it is profitable or not for them). The simulation shows that the number of adapting communities is increasing; meanwhile the number of arguing communities is decreasing. However, the number of careless communities is increasing in the first half of simulation and starts decreasing in the second half of simulation. According to this result, communities socially interact with and influence each other because in the first half, the number of arguing communities was larger than that of adapting communities, which affects the increasing number of careless communities, and then in the second half, the number of adapting communities was larger than that of arguing communities, which influence the decreasing number of

N.F. Pambudi (✉)
Graduate student in School of Business and Management Institut Teknologi Bandung, Bandung, Indonesia
e-mail: noorhan.firdaus@sbm-itb.ac.id

A. Adhiutama (✉)
Lecturer in School of Business and Management Institut Teknologi Bandung, Bandung, Indonesia
e-mail: akbar@sbm-itb.ac.id

U.S. Putro et al. (eds.), *Agent-Based Approaches in Economics and Social Complex Systems IX*, Agent-Based Social Systems 15, DOI 10.1007/978-981-10-3662-0_16

careless communities. It also shows that the role of government to maintain policies profitable for community is important.

Keywords Agent-based modeling • Community • Government • Household waste management system • Participatory action research • Waste bank

1 Introduction

Increasing population in Indonesia has led to increasing amount of waste generated by the community. Indonesian government starts to find a way in order to decrease the amount of waste which are sent to sanitary landfill. It is because sanitary landfill usage will decrease the amount of land area in one district (we consider it as district since waste management policy is conducted by district government, not national government). One way to do that is by increasing the number of waste recycled, reduced, reused, and composted in a household level.

Indonesian government through Environmental Ministry had introduced waste bank to increase recycle, reduce, and reuse activities in community. Waste bank is one policy from the government which lets the community turn their waste into money in the form of savings like in a commercial bank. But to conduct this policy, the government needs community involvement, support, and willingness to exchange their waste. This policy already showed the impact into waste tonnage which was sent to landfill in Surabaya city. Previous research in Surabaya indicated that the reduction of waste tonnage up to 7.14 tons per week was supported by the increasing number of waste bank in Surabaya by more than 50 and 30% in 2012 and 2013 [16].

The processes which are involved in this research include the waste collecting phase in the community where wastes are sold to specific/contracted small and medium enterprises or remanufacturing enterprises. There are several actors which will be observed in this research such as the community, government, waste bank administrators, and remanufacturing industries. This research aims to define behavior in each actors and its impact to performance of household waste management system.

Agent-based modeling is chosen to describe interaction among communities and government in order to gain community involvement in household waste management system. Influencing factors will also be added to the model to see how these factors will impact the dynamic of community and influence it to change its behavior, from not caring too much about their environment, especially waste, to caring about environment and adopting the policy given by government.

2 Literature Review

2.1 Closed-Loop Supply Chain in Household Waste Management System

Reverse logistic is the process of planning, implementing, and controlling the efficient, cost-effective flow of raw materials, in-process inventory, finished goods, and related information from the point of consumption to the point of origin for the purpose of recapturing value or proper disposal [11]. A forward supply chain is a network of facilities and distribution options that performs the function of procurement of materials, the transformation of these materials into intermediate and finished products, and the distribution of these finished products to customers [8]. Composting and recycling system often occurred in reverse logistic of supply chain management. But the processes of producing packaging material in several kinds of products are involved in forward logistic and also the processes to recycling waste material to a better value of products. Closed-loop supply chain is a combination of reverse logistic and forward logistic [8].

Although collection, recycling, and disposal procedures for used and obsolete products are important components of corporate responsibility [5], collection, recycling, and disposal processes are also conducted by a third party such as scavengers in developing countries such as Indonesia. Indonesian government through regulation [10] has introduced a recycling system in household level through waste bank policies. Waste bank is a place for community to exchange their waste into money in the form of savings like in a bank. Waste bank is considered to be involved in closed-loop supply chain in household waste management system. It can be seen in Fig. 1.

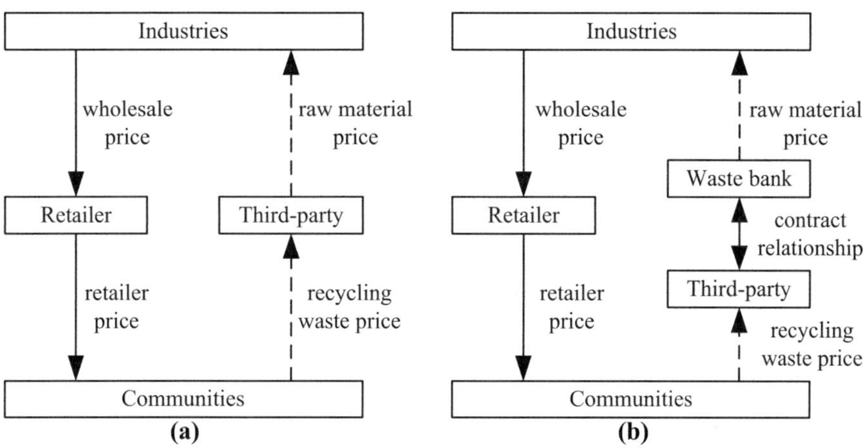

Fig. 1 Closed-loop supply chain without waste bank (**a**) and with waste bank (**b**)

In several researches, closed-loop supply chain problem was solved by several types of method. Retailer and non-retailer models of closed-loop supply chain in electronic industries are solved by defining constraints in retailer and non-retailer model [5]. A genetic algorithm approach also used in previous research for solving a closed-loop supply chain in battery recycling industry [8]. But the problem in firms' environmental behavior (dynamic of environmental behavior in Chinese firms) is solved by agent-based modeling approach [9]. The behavior in each part of closed-loop supply chain can be approximated by agent-based modeling; however, the closed-loop supply chain pricing and costing should be approximated by mathematical modeling or heuristic method such as genetic algorithm.

2.2 *Waste Management System in Indonesia*

Problems faced in waste management system [1]:

- Equipment capacity is not suitable.
- Equipment are less maintained.
- Development for operational personnel especially for daily worker is weak.
- Operational method suitable for each area condition is limited.
- Operational cycle in waste management is not completed because of different persons who are responsible.
- Coordination among departments in government is usually weak.
- Operational management is more focused on implementation, but in the controlling aspect, it is weak.
- Operational planning is often only used for short-term period.

In Indonesia, through Peraturan Menteri Lingkungan Hidup No. 13 Tahun 2012 about guidelines for implementation in reduce, reuse, and recycle with bank sampah (waste bank), activities for decreasing nonorganic waste are done with waste bank system. In that regulation, waste bank system that dedicated a place for separating and collecting wastes which can be recycled and have economic values had been introduced. With waste bank, societies are able to exchange their waste for a specific type which can be recycled and/or reused with money in savings that can be used in specified term. Price of waste in this regulation considered environmental cost from production processes for recycling waste into specified product until it have no economic value at all.

Current problems that happened in waste bank implementation is there are scavengers or other parties outside waste bank that doing similar activities and couldn't be controlled by government in several areas. Otherwise, the number of waste banks can't accept needs for recycling and reusing activities in some place. This regulation explained that market price is fluctuating, which affects the change in price according to market needs and also influences customers to sell their waste to buyers who offer higher price.

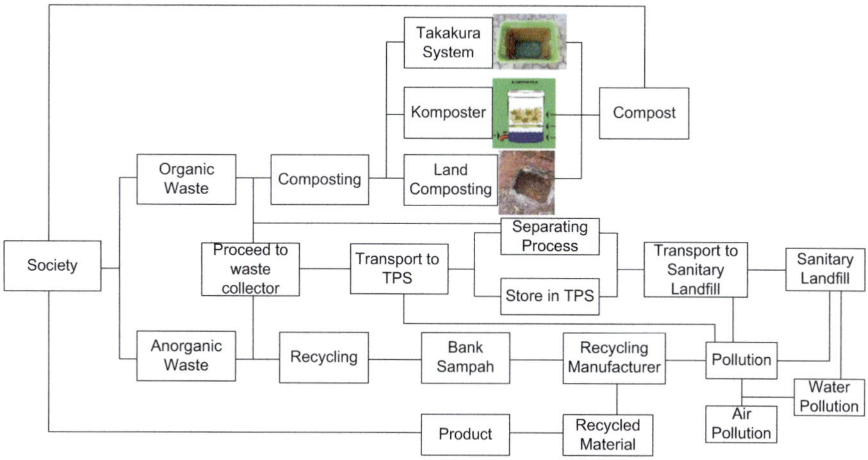

Fig. 2 Waste management system that considers waste bank and three ways of composting

This difference among the prices will affect the competition between waste bank and scavengers. Savings system that has a goal of decreasing consumptive behavior of societies can't be achieved if customers sell their waste to scavengers rather than waste bank. For that reason, societies have a decisive role in this situation for giving more integrated and sustainable system for reduce, reuse, and recycle implementation in household waste management.

Besides implementation of waste bank, government also introduced several systems that are able to support composting activities at a household level. There are three systems that are introduced by the Ministry of Environment, including Takakura system, simple composter, and composting activities, that use empty space or land in the household (this system can be done if household has an empty space). Unfortunately, those three systems can only be implemented well if there is no active participation from societies. Figure 2 illustrates the process of waste management system if all systems introduced by government that support reuse, reduce, and recycle activities can be run well.

In several researches, waste management problems were solved by several types of method. Problem in waste paper procurement optimization is solved by agent-based simulation approach [12].

3 State of the Art

Several researches have been studied about agent-based modeling in environmental aspect, such as the firms' environmental behavior [9], taxonomy of agent-based modeling in environmental management [4], modeling personality and power as evidence is brokered to support decisions on environmental risk [2], waste paper

Table 1 State of the art of this research

Author	Object	ABM	Waste management	Environmental aspect	Closed-loop supply chain	Heuristic method (LP)
[3]	Landfill construction in Hong Kong	X	X	X		
[4]	Taxonomy of ABM in environmental management	X		X		
[9]	Firms' environmental behavior in China	X		X		
[2]	Several case studies about regulation	X		X		
[12]	Waste paper procurement	X	X			X
[13]	Solid waste management in Florida, USA	X	X	X		
[5]	Electronic industries		X	X	X	X
[8]	Battery recycling		X	X	X	X
SOTA	Recycling system (waste bank) in household waste management system	X	X	X	X	Participatory action research

procurement optimization [12], single-stream recycling programs [13], and backfill in construction waste management (Gan and Cheng). Each research develops different agents and behaviors. Firms' behavior when facing environmental regulation [9] is focused on, as well as regulators or government that controls environmental risk [2]. Recycled product industries also become an agent of waste paper procurement optimization [12], as presented in Table 1.

4 Method

4.1 Participatory Action Research

In this subchapter we talk about participatory design and participatory action research. Several studies talked about participatory design and participatory action research separately. It is actually already common sense to use the terminology [15]. Designers and final users will learn from each other through feedback. Designers will lead the designing process in the initial step, giving a clear instruction and explanation on the objectives from interaction between designers and users. Clear objectives will lead the designers to focus on results, a pattern which can be formed as alternatives of solution for problems that occurred. In PD, users and designers should be in the same position, and interventions are not allowed in generating idea process.

A lot of constraints associated with human aspects; social, culture, and religious aspects; financial and time frame aspects; and organizational aspects are faced when using PD for designing an exact product for societies (6). Most research that employs PD as its methodology discusses about discrimination issues in societies which affect opinions or unrecognized needs of several elements of societies. Discriminations on people with disabilities in developing country like Cambodia (6) and disparities in health status because of their economic class and the racial differences between white and nonwhite in the USA exist [7, 14].

Participatory design that was used in that condition is caused by this methodology giving freedom in argument for several parties who are not given an opportunity in giving their aspiration. Wanyama and Zheng [15] called it as "democracy in the workplace." PD not only gives an opportunity to marginalize or discriminate community but also can be used to give an opportunity to several parties who are not capable to develop a product but have an influence in deciding available needs in societies. Those parties are consumers of the products. Previously, PD also was used for developing digital products in an information system. Developers of IT products are well educated in computer system, but they have a limitation for making an understandable system for its consumers. PD appeared to give a description of IT products that can be used by its consumers easily [15].

Hussain et al. [6] repaired tradition system of PD by changing it into an integrated methodology as seen in Fig. 2. Previously, PD conducted with only solved designers and users in one term without any specific division of role for avoiding designer's intervention to users' idea and vice versa. Especially with the appearance of stakeholders as one part that should be needed to be involved, the previous form of PD is not strong enough to produce a solution. Detailed explanation about what kind of necessary interaction happened between designers, end users, and stakeholders is needed.

4.2 Agent-Based Modeling and Simulation

4.2.1 Purpose

The purpose of this model is to explore community involvement regarding their behavior in adopting composting and recycling systems (waste bank) which are proposed by the government. The model also examines the dynamic of different perspectives from community and government to adopt and publish policies in household waste management.

Table 2 Overview of state variables and scales

Variables	Description
Numbers of common community (agents)	The amount of common community
Numbers of careless community	The amount of careless community which will be rejected from the system
Numbers of arguing community	The amount of community which always argues about policy
Numbers of adapting community	The amount of community which adapts the policy
Rate of careless community	Ratio between numbers of careless community and numbers of common community
Rate of arguing community	Ratio between numbers of arguing community and numbers of common community
Rate of adapting community	Ratio between numbers of adapting community and numbers of common community
Careless community	Community or agents who are not willing to adapt the household waste system (waste bank)
Arguing community	Community or agents who are still arguing price and cost comparison; they are willing to adapt the household waste system if there is "cheaper" condition
Adapting community	Community or agents who are willing to adapt whether price and cost condition is not "cheaper"

4.2.2 Variable and Overview of the Model

A community and government are characterized by state variables. The variables selected are substantiated by the literature or a field survey during participatory design. Variables included in this research are listed in Table 2, with two considerations on simulation to decide which transformation will be taken by the community. The first consideration is *price and cost comparison*, i.e., whether the price is higher or lesser than the cost. If the price is higher than the cost, this condition is referred to as the "cheaper" condition. The second consideration is *willingness of community* to adopt waste bank (household waste system). If the community is willing to adopt waste bank, then the condition is termed the "willing" condition.

Initial condition of the community is called the common community. Numbers of common community or agents are 100 which will be transformed to adapting community, arguing community, or careless community. Its transformation depends on the condition of each consideration (price and cost comparison and willingness of community). If the iteration has a "willing" condition whether it has a "cheaper" condition or not, then the community will be transformed to adapting community. If the iteration has no "willing" condition because it has no "cheaper" condition, then the community will be transformed to arguing community. If the iteration has no "willing" condition but it has "cheaper" condition, then the community will be transformed to careless community. It was described in Fig. 3. Price is represented by letter "p" in the figure, cost is represented by letter "c," and threshold "p>c" means whether the condition is "cheaper" or not.

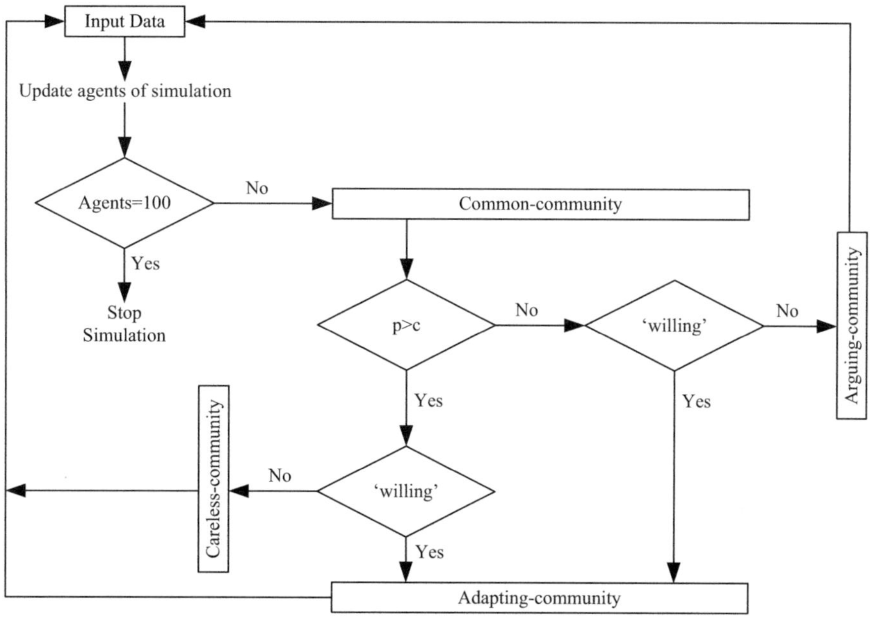

Fig. 3 Flowcharts of the model's processes

4.2.3 Model Description

This research scope is household waste management system which consists of interaction between communities and government (policy maker). Communities will accept the policy in shape of system in recycling and composting strategies in household level. Waste bank will be considered as one policy introduced by government in this model. The interaction will be dependent on type of tool provider (government or communities itself provide the composting tool), waste bank development (number and location of waste bank built in district), third-party involvement (availability of third party that maintains waste collection in household, i.e., scavengers or cleaning service paid by communities), community pressure (pressure from community to other community and government to repair their way of thinking or policies that will be made), policy pressure (pressure from policy that makes communities obey it in order to avoid sanction), the number of autocracy policy (whether a lot of policy did not consider communities' needs), and the number of democracy policy (whether a lot of policy have been considering communities' needs).

The model wants to capturing a changing behavior of communities and government in order to maintain their own perspective about the policy or household waste management system. Community can be moved from common communities (who just know the information about the policy) to adapting communities (who are already willing to adopt the policy) or to arguing communities (who will think about the impact of policy once again before deciding) or to be rejected from the model and become a careless community who doesn't care about the policy at all. The government can change their way to produce policy by increasing or decreasing autocracy policies or increasing or decreasing democracy policies. It will be the impact to communities' perspective, and this is actually the interaction among communities and government.

4.2.3.1 Collection Phase in Household

There are several options in conducting collection phase in household waste management system. There is hiring third party (i.e., scavengers or cleaning services) who will be paid periodically by communities or no hiring third party which means communities will send the waste to waste bank or temporary final disposal site by themselves. But if waste bank is developed very well until the number of waste bank increases and the location of waste bank is possible to hiring waste bank administrators to collect our waste periodically, then there will be another option to hiring waste bank administrator in collection phase. If communities hire third party, then they will not accept the advantages of using waste bank, and in this option, communities still have an opportunity to make savings in waste bank; however, exchange value must be subtracted by collection cost previously. This will be influencing to cost in collection waste and will be considered as situation in changing behavior of communities.

4.2.3.2 Exchange Phase from Household to Waste Bank

There are several types of conducting exchange phase from household to waste bank. It will be connected to collection phase. If in the collection phase communities use third party to collecting waste, then there is no advantage that can be achieved by using waste bank. If in the collection phase communities exchange their waste independently, then they will receive advantages of savings in waste bank. If in the collection phase communities use waste bank administrator to collect their waste and exchange it, then communities will receive advantages minus their responsibility to pay waste bank administrator (collection cost using waste bank administrator services).

4.2.3.3 Collection Phase in Waste Bank

Waste bank administrators must differ waste into specific waste in their own characteristic such as plastic waste, metal waste, paper waste, etc. These activities can be conducted when waste bank administrator collects waste in household (it happened if household uses waste bank administrator services to collect their waste) and in waste bank site. This activity will give production lead time until waste is ready to be sent to industries (small medium enterprise or recycling industries).

4.2.3.4 Selling to Industries

After the collecting phase in waste bank, waste bank administrator will sell waste to industries. Specific price will be taken for a specific type of waste. This activity will give profit and ability to pay communities for waste bank. The profit can be used as developing ability of waste bank to build another waste bank. However, building another waste bank in other location will also be government's responsibility.

4.2.3.5 Inspection from the Government About Waste Bank

Waste bank will be inspected by the government periodically, not only waste bank performance but also whole of supply chain performances in composting and recycling system in household waste management system. It is because performance of each part of supply chain such as communities, third parties, recycling and small medium enterprise industries, local government, and waste bank administrators will impact other part of supply chain. It will impact waste bank development in other local areas if there is some success story resulted from waste bank system in another local area. This experience also will be adopted by other waste bank in other local area. It is similar with technology commercialization scheme. That's the reason why agent-based modeling is appropriate for this study.

5 Results and Discussions

The model will be simulated and compared with different situation of the system. Types of situations are considered waste bank in household waste management system, household waste management system without waste bank, increasing number of waste bank and location of waste bank, and different types of variables mentioned in Sect. 4. From that comparison, this research offered several actions that must be taken in different situation, what should government do to maintain household waste management system and how the relationship between government and communities in household waste management system.

Further research can be considered about other policy besides waste bank or adopting the model with different policies based on different countries or governments. Schematic of agent-based modeling can be combined with other methods in simulation such as system dynamics or linear programming to determining cost or price which has become data input in agent-based simulation.

In this model we assumed that there are two considerations in adapting waste bank by households. There are price and cost comparison and willingness from

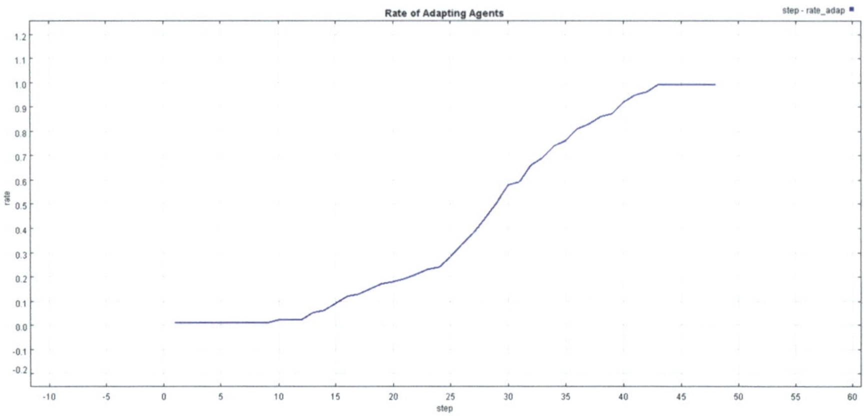

Fig. 4 Results of the simulation and rate of adapting community

community. If the price (selling price of waste exchange) is higher than cost (transportation and collection cost), then we called it as "cheaper" condition in the simulation. If there is willingness from household to use waste bank, then we called it as household "willing" to use waste bank whatever condition from price and cost comparison. The simulation used SOARS 4.1.1 to describe the result of the simulation. We want to know the rate of common community who transforms to become adapting community.

Figure 4 showed the rate of adapting community from the result of the simulation. It described that the rate of adapting community is increasing. In the beginning of simulation, the rate of adapting community is zero which means that there is no community adapting the household waste system (waste bank). Then in the end of simulation, the rate becomes one which means that all members of community or all agents have adopted the system. This increase happened because there is changing in price and cost which influences changing in "cheaper" condition.

Figure 5 showed the rate of arguing community from the result of the simulation. It described that the rate of arguing community is decreasing. In the beginning of simulation, the rate of arguing community is one which means that all members of community or all agents are arguing about the system. Then in the end of simulation, the rate becomes zero which means that there are no members of community or agents who are arguing about the system. This decrease happened because of the impact from increasing pattern for the rate of adapting community.

Figure 6 showed the rate of careless community from the result of the simulation. It described that the rate of careless community is increasing in the first half of iterations and then decreasing in the second half of iterations. This pattern happened because careless community is the transformation when the community has no "willing" in the "cheaper" condition. For "willing" condition represented by condition in the rate of adapting community because adapting community is community who always "willing" whether there is no "cheaper" condition. For

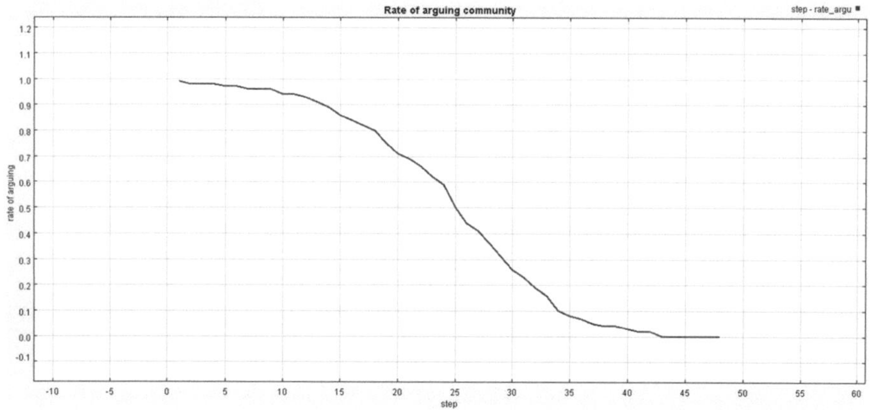

Fig. 5 Results of the simulation and rate of arguing community

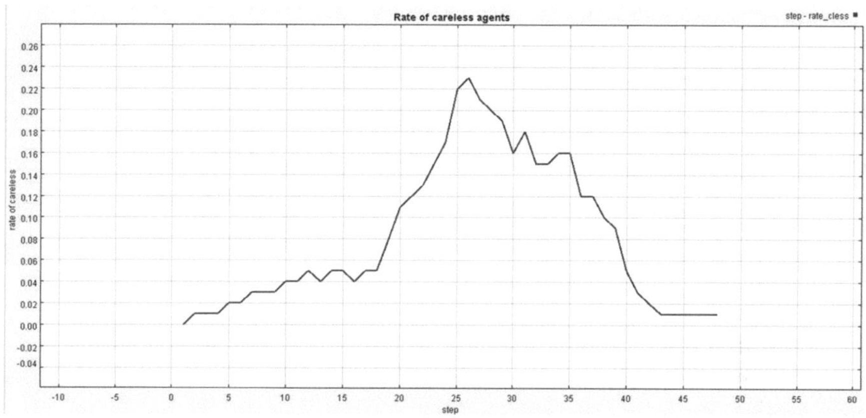

Fig. 6 Results of the simulation and rate of careless community

"cheaper" condition represented by condition in rate of arguing community because arguing community is community who has no "cheaper" condition and decide to have no "willing" before there is "cheaper" condition. In the first half of iterations, the numbers of arguing community are higher than the numbers of adapting community which made the increase in the numbers of careless community. In the second half of iterations, the number of adapting community is higher than arguing community which made the decrease in the numbers of careless community. It was shown in Fig. 7.

Figure 7 showed that when the rate of adapting community is increasing but the number of adapting community is less than the numbers of arguing community, then rate of careless community will be increasing. It is because "willing" condition is still dominated by no ("willing" = "no") value. But when the graphs are much closest to equilibrium point between the rate of arguing community and

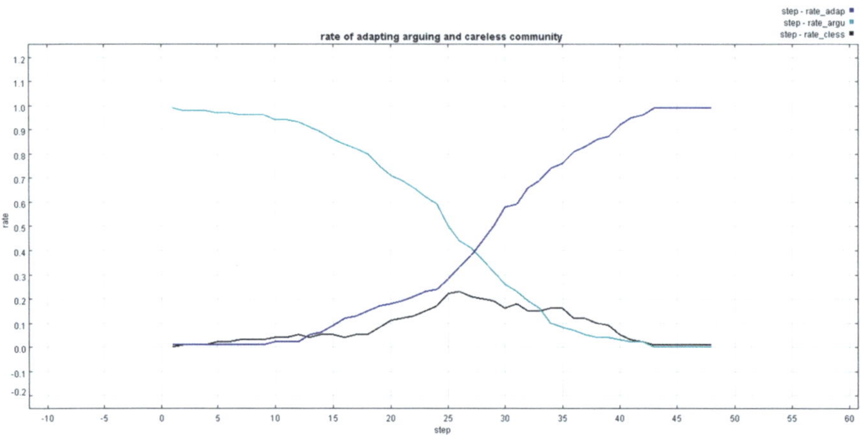

Fig. 7 Comparison between the rate of adapting community (*blue*), arguing community (*green*), and careless community (*black*)

rate of adapting community, there is a decreasing pattern from the rate of careless community although the numbers of adapting community are less than the numbers of arguing community. It means that "willing" condition is no longer dominated by no ("willing" = "no") value. The decrease becomes tight when the numbers of adapting community are much higher than the numbers of arguing community.

6 Conclusions

Price and cost which are represented by "cheaper" condition became an influencing factor besides the willingness of community to adopt the system (waste bank). Decreasing cost can be maintained by developing several policies which can eliminate transportation cost and collection cost. The role of government to provide the tools in household level to make an ease in the collecting process can eliminate collection cost. Policy to provide service which delivers waste from household to waste bank will eliminate transportation cost. This kind of policy can make cost near to zero that affects the increase of "cheaper" condition. It can be developed with the willingness of community to use waste bank rapidly, but willingness of community also must be our concern to apply waste bank in community. It is happened because there are conditions that community will transform to careless community who will have no "willing" whether the condition of price and cost comparison is "cheaper" (price is higher than cost). Further research can be considered about other policy besides waste bank or adopting the model with different policies based on different countries or governments. Schematic of agent-based modeling can be combined with other methods in simulation such as system dynamics or linear programming to determine cost or price which has become data input in agent-based simulation.

Other qualitative approach also can be considered to combine with modeling to give an insight about decision based on emotional and psychological aspect of human being. Unfortunately, this research more focuses on simulation rather than discusses about the result from participatory design or participatory action research.

From the result itself, this research proposes to maintain the number of adapting community larger than the number of arguing community. To maintain the number of arguing community, it will need more contribution from government to pay much attention in order to give more profitable condition for community when they try to adopt eco-friendly policies. This condition shows that collaboration between government and community will be needed and necessary for the future. It also gives the idea to get the research which will accommodate government and community conflict of interest. It means combination of two approaches, bottom-up approaches where ideas are generated by the community and top-down approaches where ideas are generated by the government. Combination of agent-based modeling and system dynamic is appropriate for this kind of research. Government also can use participatory action research to design new policies and technologies which are appropriate for community. Combination of design thinking and system thinking will be necessary to this idea.

References

1. Damanhuri, E., & Padmi, T. (2000, October). Reuse and recycling as a solution to urban solid waste problems in Indonesia. In *Proceeding of the international symposium and exhibition on waste management in Asian Cities*, Hong Kong.
2. Davies, G. J., Kendall, G., Soane, E., Li, J., Rocks, S. A., Jude, S. R., & Pollard, S. J. (2014). Regulators as agents: Modelling personality and power as evidence is brokered to support decisions on environmental risk. *Science of the Total Environment, 466*, 74–83.
3. Gan, V. J., & Cheng, J. C. (2015). Formulation and analysis of dynamic supply chain of backfill in construction waste management using agent-based modeling. *Advanced Engineering Informatics, 29*, 878–888.
4. Hare, M., & Deadman, P. (2004). Further towards a taxonomy of agent-based simulation models in environmental management. *Mathematics and Computers in Simulation, 64*(1), 25–40.
5. Hong, I. H., & Yeh, J. S. (2012). Modeling closed-loop supply chains in the electronics industry: A retailer collection application. *Transportation Research Part E: Logistics and Transportation Review, 48*(4), 817–829.
6. Hussain, S., Sanders, E. B. N., & Steinert, M. (2012). Participatory design with marginalized people in developing countries: Challenges and opportunities experienced in a field study in Cambodia. *International Journal of Design, 6*(2), 91–109.
7. Israel, B. A., Coombe, C. M., Cheezum, R. R., Schulz, A. J., McGranaghan, R. J., Lichtenstein, R., ... & Burris, A. (2010). Community-based participatory research: A capacity-building approach for policy advocacy aimed at eliminating health disparities. *American Journal of Public Health, 100*(11), 2094–2102.
8. Kannan, G., Sasikumar, P., & Devika, K. (2010). A genetic algorithm approach for solving a closed loop supply chain model: A case of battery recycling. *Applied Mathematical Modelling, 34*(3), 655–670.

9. Liu, Y., & Ye, H. (2012). The dynamic study on firm's environmental behavior and influencing factors: An adaptive agent-based modeling approach. *Journal of Cleaner Production, 37*, 278–287.
10. Republik, P. M. N. L. H. (2012) Indonesia Nomor 13 Tahun 2012 TentangPedomanPelaksanaan Reduce. *Reuse, dan Recycle melalui Bank Sampah.*
11. Rogers, D. S., & Tibben-Lembke, R. S. (1999). *Going backwards: Reverse logistics trends and practices*. Pittsburgh: RLEC Press.
12. Sauvageau, G., & Frayret, J. M. (2015). Waste paper procurement optimization: An agent-based simulation approach. *European Journal of Operational Research, 242*(3), 987–998.
13. Shi, X., Thanos, A. E., & Celik, N. (2014). Multi-objective agent-based modeling of single-stream recycling programs. *Resources, Conservation and Recycling, 92*, 190–205.
14. Wallerstein, N., & Duran, B. (2010). Community-based participatory research contributions to intervention research: The intersection of science and practice to improve health equity. *American Journal of Public Health, 100*(S1), S40–S46.
15. Wanyama, I., & Zheng, Q. I. N. (2010). Augmenting the participatory design concept in systems development. *Management Science and Engineering, 4*(2), 69–81.
16. Wijayanti, D. R., & Suryani, S. (2015). Waste bank as community-based environmental governance: A lesson learned from Surabaya. *Procedia – Social and Behavioral Sciences, 184*, 171–179.

MIX
Papier aus verantwortungsvollen Quellen
Paper from responsible sources
FSC® C105338

If you have any concerns about our products,
you can contact us on
ProductSafety@springernature.com

In case Publisher is established outside the EU,
the EU authorized representative is:
Springer Nature Customer Service Center GmbH
Europaplatz 3, 69115 Heidelberg, Germany

Printed by Libri Plureos GmbH
in Hamburg, Germany